Mind World

Essays in Phenomenology and Ontology

This collection explores the structure of consciousness and its place in the world, or inversely the structure of the world and the place of consciousness in it. Among the topics covered are the phenomenological aspects of experience (inner awareness, self-awareness), dependencies between experience and the world (the role of the body in experience, the role of culturally formed background ideas), and the basic ontological categories found in the world at large (unity, state-of-affairs, connectedness, dependence, and intentionality). Developing ideas drawn from historical figures such as Descartes, Husserl, Aristotle, and Whitehead, the essays together demonstrate the interdependence of ontology and phenomenology and its significance for the philosophy of mind.

David Woodruff Smith is Professor of Philosophy at the University of California, Irvine.

Mind World

Essays in Phenomenology and Ontology

DAVID WOODRUFF SMITH

University of California, Irvine

CAMBRIDGE
UNIVERSITY PRESS

CAMBRIDGE
UNIVERSITY PRESS

University Printing House, Cambridge CB2 8BS, United Kingdom

One Liberty Plaza, 20th Floor, New York, NY 10006, USA

477 Williamstown Road, Port Melbourne, VIC 3207, Australia

314-321, 3rd Floor, Plot 3, Splendor Forum, Jasola District Centre, New Delhi - 110025, India

79 Anson Road, #06-04/06, Singapore 079906

Cambridge University Press is part of the University of Cambridge.

It furthers the University's mission by disseminating knowledge in the pursuit of education, learning and research at the highest international levels of excellence.

www.cambridge.org
Information on this title: www.cambridge.org/9780521539739
© David Woodruff Smith 2004

First published 2004

A catalogue record for this publication is available from the British Library

Library of Congress Cataloging in Publication data
Smith, David Woodruff, 1944–
    Mind world : essays in phenomenology and ontology / David Woodruff Smith.
        p.   cm.
    Includes bibliographical references and index.
    Contents: Three facets of consciousness – The cogito circa A.D. 2000 – Return to consciousness – Consciousness in action – Background ideas – Intentionality naturalized? – Consciousness and actuality – Basic categories – The beetle in the box.
    ISBN 0-521-83203-9 – ISBN 0-521-53973-0 (pbk.)
    1. Consciousness.   2. Phenomenology.   3. Ontology.   I. Title.
B808.9.S63   2004
126 – dc22   2003061668

ISBN 978-0-521-53973-9 Paperback

For Mary and Wyndham,
whose creative minds complete my world

Contents

Prolegomena: The *Terroir* of Consciousness and the World

This book explores the structure of consciousness and its place in the world or, inversely, the structure of the world and the place of consciousness in it. Some essays focus on phenomenological aspects of conscious experience; others on the world at large, especially the importance of basic ontological categories. Some develop ideas drawn from historical figures (e.g., Descartes, Husserl, Aristotle, Whitehead), while looking to structures of consciousness and the world. I want to put these essays between the covers of one book because, as I see them, their views work together like photos of a common field taken from different perspectives.

The ideas gathered here have evolved mostly in the *terroir* of California, where ideas and cultures mix uncommonly. California phenomenology. California ontology. California syncretism. Not without a sense of history (even in California).

The essays cut across the fields of phenomenology and ontology. The interdependence of ontology and phenomenology, as well as its significance for philosophy of mind, is a running theme of the collection as a whole. This interdependence I see as part of the systematic character of philosophy as a whole, a systematic unity rejected by much of twentieth-century philosophy, not least in separating phenomenology from wider metaphysics (in the wake of Kantianism, positivism, pragmatism, existentialism).

Frequently the essays address issues in philosophy of mind, the most vigorous area of recent philosophy in the analytic tradition. Yet my perspective does not begin with current issues of the relation between mind and body – issues of physicalism, functionalism, supervenience, and the like (as in Fodor, Dennett, Searle, Dretske, Kim, the Churchlands).

Rather, my perspective begins with more purely phenomenological and ontological issues: issues of consciousness, intentionality, and ontological categories.

The essay in Chapter 1 distinguishes three fundamentally different ontological aspects of consciousness, which separate areas of terrain explored variously in other essays. Subsequent chapters analyze consciousness in its basic phenomenological structure (intentionality, inner awareness, volition, action), in its environmental conditions (brain activity, physical surroundings, cultural background), and ultimately in its "deep" ontological structure (basic categorial forms or modes of being, such as being intentional, being dependent on brain, culture, etc.). The later chapters explore ontological categories in their own right, while keeping an eye on consciousness.

The collection affords, I hope, a unified though by no means complete view of mind-and-world (the unity implied in the book's title). However, I do not wish to axiomatize here; I let the essays speak for themselves after previewing them with a broad story line. I find the essays mutually reinforcing, although each was written to stand alone, as far as possible. There are points of overlap, interlocking the essays, yet common themes may be approached from different directions.

The essays are broadly analytic in style and approach (Austro-Anglo-American philosophy). They are often phenomenological in content and method and background (continental European philosophy). They are recurrently ontological or metaphysical in content (joining a long history of Western philosophy). And they are sometimes historical in content and method (reflecting a long and global tradition of philosophy). In this way the book combines elements of philosophy that are often kept separate.

There is a vision that I hope develops over the course of the present book: that the structure of consciousness, with all the properties we find in it through phenomenology, finds its home in the complex structure of the world, with all the forms we find recounted in a systematic ontology. Only by working our way into both phenomenology and ontology, in an integrative way, can we develop this vision.

The term *terroir* is used by French vintners to incorporate all the elements of the vineyard, including the roles of geology, climate, and culture in the making of wine. The term has caught on in the California wine country. Now, the philosophical *terroir* of California has produced local varietals of phenomenology and ontology, and these essays partake of those varietals in seeking the *terroir* of consciousness in the world.

Consciousness: its distinctive experiential characters, its place in the ground of things, its being in the air in philosophy, its cultural inheritance – these are the things we must attend to in consciousness. The world: its basic categories, its diverse formal structures, its niche for consciousness – these are the things we must attend to in the world at large.

How shall we study structures of consciousness and the world?

We understand consciousness in the first place by simply experiencing it. Phenomenology is the philosophical discipline that seeks to describe, interpret, and analyze our own conscious experience, just as we experience it from our own first-person perspective. But phenomenology alone does not tell us the place of consciousness in the world.

We understand the world at large by way of everyday experience and increasingly by what empirical science discovers and hypothesizes. Ontology, or metaphysics, is the philosophical discipline that seeks to analyze the basic shape of the world. But today ontology is accountable to the remarkable empirical-theoretical results of modern science. Accordingly, the task of ontology is to frame what we know about the world, including structures of our own conscious experience, structures of the things we encounter in everyday life, and structures of what we know through empirical science.

Interestingly, the approach of modern science has left consciousness itself as one of the most pressing problems of cognitive neuroscience (the *science* of mind) and the hottest topic in recent philosophy of mind *cum* cognitive science. We philosophers and scientists together have not found a way to fit the properties of consciousness as we experience it (featuring intentionality, inner awareness, sensory qualia, volition, etc.) into the world as we know it in contemporary science (featuring quarks, quantum fields, evolution of organisms, black holes, etc.). In order to understand the nature of consciousness and how it fits into the world, what we need from the side of philosophy is a more careful synthesis of more careful phenomenology and more careful ontology. Yet the disciplines of phenomenology and (formal) ontology are too little evident in the exhilarating discipline today called philosophy of mind or, in its more scientific reaches, cognitive science. Indeed, the thought of bringing together all these fractious disciplines (and their disciples!) calls to mind an old Bette Davis line: "Fasten your seat belts, . . . it's going to be a *bumpy* night!" Nonetheless, that is what lies ahead of us.[1]

In the following chapters I pursue a particular synthesis of phenomenology and ontology. However, I leave for the Appendix a more explicit account of how I conceive of phenomenology and ontology and their integration. These are matters of controversy, reaching into broad metaphilosophical positions; I try to stake out my territory in these matters in the Appendix.

My conception of phenomenology and ontology and their interdependence, as practiced here, has evolved through several overlapping "eras" in my philosophical experience. A sense of this background may help to indicate where I am going as well as where I am coming from, and this sense calls to mind my gratitude to a number of teachers (including my students and collaborators).

When I moved from mathematics into philosophy as a graduate student at Stanford in the late 1960s, what crossed my path, after Donald Davidson's philosophy of language, was Jaakko Hintikka's possible-worlds logic of perception followed by Dagfinn Føllesdal's development of Husserlian phenomenology. By 1982 Ronald McIntyre and I had put together our collaborative interpretation of these things in a book that took us ten years to write. Here was California phenomenology at work.[2]

At Stanford I also acquired an appreciation of the history of philosophy and its contemporary relevance, absorbed from the teaching of John D. Goheen and two European logicans (Dagfinn and Jaakko) who read historical texts as seriously as today's. When I took my first philosophy courses, at Northwestern while an engineering and mathematics student, I wanted to know the truth about things, not who said it and when. Slowly, but surely, I have learned that the genealogy of philosophical concepts, in texts of bygone years, carries a great deal of their content. Indeed, there is something phenomenological in understanding an idea by tracking its historical evolution. With ideas as with organisms, ontogeny recapitulates phylogeny. In William Faulkner's words, "The past is not forgotten, it is not even past."

In 1982 I got a telephone call inviting me to talk about phenomenology's relevance to software design. Thus began my twenty-year philosophical discussion with Charles W. Dement. We started by assessing Sartre's ontology, and we have been talking ever since about issues in formal ontology *cum* phenomenology, ranging from Anaximander and Aristotle to Husserl, Ingarden, Whitehead, and Ernst Mayr (yes, the biologist). This work was part of a formal research program at Ontek Corporation (incorporated in 1985): designing systems of computational ontology *cum* phenomenology. (Ontek's work was the first of this kind.) Here was California ontology in the making. In the late 1980s and throughout the 1990s, our philosophical research – and the iterations of systems built by the Ontek team that were based on that research – moved from categories, states of affairs, and modalities (ontic and epistemic) to basic "modes" (including manyness, composition, dependence, intentionality) and on to what we have called "systematics" or "metasystematics" (marking ontological distinctions and their role in the formal genesis of entities, in a kind of analogy with biological systematics). Our work over the years also involved close collaboration with logician Peter Woodruff and ontologist Peter Simons (thinkers from the world of academic philosophy) and with Ontek systems designer Steve DeWitt and master programmer John Stanley (philosophical thinkers from the real world).

On a parallel track through the 1990s I was drawn back into Husserl's *Logical Investigations,* this time with my eyes focused more closely on formal ontology. Although my results and stance in these essays are often far from Husserlian, I found in Husserl an important case study in how phenomenology, ontology, and logic (think of formal semantics) work together. In my current view, Husserl joins Aristotle, Kant, and Whitehead as the most systematic of philosophers. A series of conferences, with occasions to present papers on these things, has helped me to develop this new perspective on Husserl and on the synthesis of ontology and phenomenology (and logic). There were conferences in Bordeaux in 1995, Leeds in 1996, Bolzano in 1997, Copenhagen in 2000, Montreal in 2001, Memphis in 2001.

For many years I have worked with small groups who gathered in southern and northern California for informal discussions of aspects of intentionality. In these forums I have talked with Ron McIntyre, Dagfinn Føllesdal, Izchak Miller (until his untimely passing), Bert Dreyfus, John Searle, Allan Casebier, Martin Schwab, Dallas Willard, Rick Tieszen, Wayne Martin, Amie Thomasson, Jeff Yoshimi, and others. I have also enjoyed conversations over the years, often in Europe, with Barry Smith, Peter Simons, and Kevin Mulligan, Ltd., the threesome British champions of formal ontology (and its history in Brentano, Husserl, et al.).

Meanwhile, I have been fortunate to work with a string of gifted graduate students at Irvine, expanding my horizons as their dissertations unfolded. In nearly weekly discussions with each, many of the ideas below have circulated in various garbs. Jeff Yoshimi, Paul Livingston, Tim Schoettle, Linda Palmer, Jason Ford, Kay Mathiesen, Joe Tougas, Dan Zelinski, Amie Thomasson, John Bickle, Jim Zaiss, Kent Baldner – these perceptive younger minds have taken very different directions. It has been fascinating to see, from my office chair, how things are related, from neuroscience to mysticism, from the ontology of art to the ontology of

politics, from phenomenology to philosophy of language to philosophy
of mind, and from one historical figure to another.

My kindest thanks to all these good people, without whom . . .

The mixing of ideas, ideals, cultures, and peoples is a natural pattern
of human evolution. California in particular presents a form of cultural
confluence demonstrating this fact of life on earth. Today's "California"
looks toward Asia and Latin America as much as Europe, as well as Africa
and the Middle East. The mix and flux of this California is a daily experience for all who have been fortunate to inhabit the richness of the
University of California in recent decades, not least on the Irvine campus
where I teach.

Yet the modern ethos – from science to human rights – is felt as deeply
threatening to many people across the globe. Science and rationality have
overridden traditions of spirituality and of ethnicity. In this century many
peoples and their values will mix more than ever before. The challenge
is how to honor difference amid sameness – not least in light of the fact
that we are all hurtling through space on a planet that we may obliterate
simply because we cannot sustain our own species along with other life
forms. The events of 11 September 2001 shook the world, in ways we have
yet to understand.

And yet, constant flux is not without history, not without continuity,
and not without form, as Whitehead's ontology declares. (The essay in
Chapter 7 draws on that ontology of flux.)

Something of the *Zeitgeist* – the constant flux of different ideas from
different cultural and philosophical origins – is at work in the essays gathered here. Not usually by design, of course, as the spirit moves us mostly
in ways we do not see. California's openness to new things has encour-
aged my looking to "other" philosophical traditions. Indeed, the mixing
of ideas in a multicultural way is an important part of the background of
the essays in this book. (Including the one titled "Background Ideas" in
Chapter 5, first published in Italian in Rome.)

My thanks thus to the intellectual *terroir* of "California."

Notes

1. The relevance of Husserlian phenomenology for cognitive science has
    been explored in two anthologies: Hubert L. Dreyfus (with Harrison Hall),
    ed., *Husserl, Intentionality and Cognitive Science* (Cambridge, Mass.: MIT
    Press, 1982); and Jean Petitot, Francisco J. Varela, Bernard Pachoud, and

Jean-Michel Roy, eds., *Naturalizing Phenomenology: Issues in Contemporary Phenomenology and Cognitive Science* (Stanford, Calif.: Stanford University Press in collaboration with Cambridge University Press, 1999). From a different perspective, the analytic tradition in philosophy of mind, in the parts most relevant to my concerns, has been highlighted in Ned Block, Owen Flanagan, and Güven Güzeldere, eds., *The Nature of Consciousness* (Cambridge, Mass.: MIT Press, 1997); however, the phenomenological tradition, with its rich panoply of results, is not drawn into the latter volume. Meanwhile, a new collection is imminent, joining results from both classical phenomenology (as a discipline) and contemporary philosophy of mind: David Woodruff Smith and Amie L. Thomasson, eds., *Phenomenology and Philosophy of Mind* (forthcoming). Of course, many analytic philosophers remain uncomfortable with the first-person, phenomenological approach to consciousness, and a few reject the existence of phenomena of consciousness. On the other hand, many phenomenologists remain uncomfortable with the naturalism of scientifically oriented philosophy of mind, often setting "transcendental" phenomenology in opposition to any naturalistic explanation of consciousness, intentional content, or meaning in general. Stay tuned for my take on such issues.

2. The school of so-called California phenomenology began with Dagfinn Føllesdal's teaching at Harvard and Stanford in the 1960s, joined by Hubert Dreyfus's teaching at Berkeley. This conception of phenomenology, and its Husserlian foundations, is laid out in two volumes: Hubert L. Dreyfus, ed., *Husserl, Intentionality and Cognitive Science* (1982, cited in note 1); and David Woodruff Smith and Ronald McIntyre, *Husserl and Intentionality: A Study of Mind, Meaning, and Language* (Dordrecht: D. Reidel, 1982). The latter work remains the only book-length development of the neo-Husserlian theory of intentionality that evolved in the "California" tradition (in fact, a particular variant on the theory emphasizing horizon as well as noema, and detailing connections with Fregean and possible-world semantic theory as well).

Origins of the Essays

The essays in Chapters 1, 2, and 4–7 were originally published as indicated and have been edited here for uniformity of style:

"Three Facets of Consciousness," *Axiomathes* 12 (2001): 55–85.

"The Cogito *circa* AD 2000," *Inquiry* 36 (3) (September 1993): 225–54.

"Consciousness in Action," *Synthese* 90 (1992): 119–43.

"Background Ideas" appeared in Italian translation as "Idee di sfondo," *Paradigmi* (Estratto da PARADIGMI, Rivista di critica filosofica) (Rome), Anno XVII, no. 49 (January–April 1999): 7–37.

"Intentionality Naturalized?" in Jean Petitot, Francisco J. Varela, Bernard Pachoud, and Jean-Michel Roy, eds., *Naturalizing Phenomenology: Issues in Contemporary Phenomenology and Cognitive Science* (Stanford, Calif.: Stanford University Press in collaboration with Cambridge University Press, 1999), pp. 83–110. The essay has been translated as "L'intentionnalité naturalisée?" in Jean Petitot, Francisco J. Varela, Bernard Pachoud, and Jean-Michel Roy, eds., *Naturaliser la phénoménologie: Essais sur la phénoménologie contemporaine et les sciences cognitives* (Paris: CNRS Editions, 2002), pp. 105–42.

"Consciousness and Actuality" appeared as "Consciousness and Actuality in Whiteheadian Ontology," in Liliana Albertazzi, ed., *The Dawn of the Cognitive Science: Early European Contributors* (Dordrecht: Kluwer Academic Publishers, 2001), pp. 269–97 – only the title is changed, for aesthetic parallel.

I am most grateful to the original publishers for permission to reprint these essays in the present book: to Taylor and Francis, Ltd., publisher of the journal *Inquiry*, in which Chapter 2 appeared; to Stanford University

Press, publisher of the volume *Naturalizing Phenomenology*, in which Chapter 6 appeared; and to Kluwer Academic Publishers, publisher of the journal *Axiomathes*, in which Chapter 1 appeared; publisher of the journal *Synthese*, in which Chapter 4 appeared; and publisher of the volume *The Dawn of Cognitive Science*, in which Chapter 7 appeared.

The essays in Chapters 3 and 8 are published here for the first time:

"Return to Consciousness": this essay aims to update, revise, and amplify my analysis of "inner awareness" in *The Circle of Acquaintance* (Dordrecht: Kluwer Academic Publishers, 1989) and "The Structure of (Self-)Consciousness," *Topoi* 5(2) (1986): 149–56.

"Basic Categories": this essay includes material from a series of four lectures I gave in the 1997 Bolzano International Schools in Cognitive Analysis at the conference "Categories: Ontological Perspectives in Knowledge Representation" in Bolzano, Italy, 15–19 September 1997; the essay reflects my work on ontology in collaboration with colleagues at Ontek Corporation since 1982 and a series of courses and seminars I taught at the University of California, Irvine, during the 1990s.

Overview: A Story Line

The Background

In a moment I will sketch a line of argument, or rather narrative, that weaves through the essays gathered here. But first let me recall some background notions broadly assumed in that story line.

Consciousness is a consciousness "of" something, and this of-ness – called *intentionality* – is the tie that binds consciousness and world together.

Intentionality is itself the structure in which we know about the world. This structure begins with mental and practical acts on the one hand and objects of various types on the other. Phenomenology works from intentionality into structures of experience, or conscious mental activity, whereas ontology works inter alia from intentionality into structures of the world in general (including mental activity). We do not normally think of ontology as beginning with intentionality. As Quine has stressed, however, our ontology consists of what we posit in our preferred theories – what we posit, I note, in our intentional activities of theorizing.

So we may think of working from intentionality into phenomenology on the one hand and into ontology on the other hand. In one direction lies "subjective" structure; in the other lies "objective" structure. Both directions are pursued in the essays gathered in this book, but the subjective and objective, I urge, are part of one world with a unified structure. (By contrast, Descartes posited two realms of mind and body, and Kant separated two spheres called phenomena and noumena, or things-as-they-appear and things-as-they-are-in-themselves.)

1

Since Husserl's work in the late nineteenth and early twentieth centuries, philosophers have come to define intentionality as the property of a mental state's being "of" or "about" something – in the sense that (following Husserl) consciousness is (almost always) a consciousness "of" something. The concept of intentionality has been developing since at least Aristotle, but it came into its own in Husserl. In the background of the essays in this volume lies a reconstruction of Husserl's basic theory of intentionality.[1] What I rely on is mostly an appreciation of the phenomenon of intentionality, including intentional content and the intentional relation of mental act to object. This much is broadly Husserlian but shared by other philosophers who take seriously "first-person" consciousness and content.[2]

Very briefly, the content theory of intentionality, in the form I prefer, holds that intentionality consists in a complex structure of context, subject, act, content, and object – that is, within a certain context a person or subject performs or experiences an act of consciousness (thinking, seeing, willing, etc.) with a certain content (thought, image, etc.) that represents or "intends" a certain object (individual, state of affairs, event, etc.). In that way consciousness is intentionally directed toward an object. Schematically:

context |—— subject — act — content ——> object.

The context includes the background conditions on which the intentionality depends. The subject is the person who is conscious. The act is the state or event or process of thinking, perceiving, imagining, desiring, willing, or whatever. The content is the ideal or abstract "meaning" entertained in the act. That content represents something, which is the object of the intentional act, that which the subject is conscious "of" – in a certain way defined by the content and conditioned by the context.

A special range of cases that have interested me are those in which the subject is directly acquainted with the object, as in visual perception. Here the content is naturally expressed by indexical words such as "this," "here," "I," "her," etc. The intentional or semantic force of the intrinsically "indexical" content of an act of acquaintance depends on the context of the act: my perception of "this" tree depends on which tree is in my visual environment as I see "this." The structure of acquaintance figures in some studies in this volume, so I point toward it here in preview. What may be less familiar is how the case of acquaintance is handled in a content theory of intentionality.[3] In this form of intentionality mind and world are most intimately connected.

The Story Line

The essays to follow tell their stories individually. But these shorter stories fit into a larger story, a broad philosophical account of mind and world. I have arranged the essays in a pedagogical order: moving mostly from more phenomenological to more ontological issues. An alternative pedagogy would move in the reverse order, and one might well read the essays in reverse, depending on one's interests. Here, in an overview, I attempt to weave the larger story line around salient themes in the individual essays.

Three Facets of Consciousness

In the information age computer scientists have found it useful to distinguish a computing system's hardware, software, and users: the *physical* implementation of the *formal* computing algorithms manipulated by humans as the computation *appears* to them on their computer screens. But this *three-schema* approach to computation reflects a highly abstract ontological framework. Indeed, the nature of any entity divides into three fundamental *facets* that we may call *form, appearance,* and *substrate.* An entity's form consists in its kinds, properties, relations; its appearance consists in the way it is known or experienced by a knowing agent; its substrate consists in that on which it depends for its existence (such as deep physical process in quarks or strings or whatever). Now, the nature of an act of consciousness divides thus into form, appearance, and substrate. Its form is intentionality; its appearance is its qualitative phenomenological character as experienced; its substrate is its neural basis, its cultural background, and more. Keeping this division of essence in mind will change the way we practice philosophy of mind and indeed ontology in general, while sharply defining the place of phenomenology in both.

The Cogito circa A.D. 2000

Philosophers have studied *intentionality,* the basic form of consciousness, in various guises at least since Aristotle. But it was Husserl's work circa 1900 that finally produced a sharp model of intentionality. On this model, an act of consciousness is directed via a conceptual structure of *meaning* (intentional content) toward an object appropriately represented or "intended" through that meaning. But how do we come to know the form of consciousness? The phenomenological turn to consciousness and its intentional structure began with Descartes's *cogito ergo sum.* The best way to appreciate the "first-person" approach to mind – which has returned to center stage in contemporary philosophical-scientific theory of mind – is to reexamine the cogito from today's perspective. Consciousness includes, in its very structure, an *inner awareness* of the transpiring act. The logic of the cogito follows this form of inner awareness. This inner awareness grounds our knowledge of our own conscious experience from our own first-person perspective. That knowledge is not incorrigible (as Ryle averred of Descartes's claim). Rather, it is the experientially certain starting point of our understanding of consciousness. And third-person studies of mind must accommodate this first-person structure.

The Return to Consciousness

What makes a mental act or state conscious, on the classical view (Descartes, Locke, Brentano, Husserl, et al.), is a certain *inner awareness* of the act as it transpires: I am not consciously thinking, perceiving, and the like unless I am aware of so thinking or perceiving. What is the *form* of that inner awareness? It cannot be that of a distinct mental act of observing or reflecting on the given mental act, because then we have two acts instead of one and tend toward an infinite regress (of observing observing . . .). Instead, inner awareness must be an integral component of a conscious experience. Roughly speaking, we may articulate the structure of inner awareness in the following form of phenomenological description: "Phenomenally in this very experience I see this frog." The inner awareness is not, then, an additional and second-order awareness but rather an integral self-reflexive component of the given act. In this way we may avoid the problems of recently fashionable "higher-order" theories of consciousness. Nonetheless, we should recognize a gradation from elementary sentient consciousness to more complex forms of consciousness, recognizing that it is these "higher" forms that involve inner awareness.

Consciousness in Action

Since Descartes's revolution, turning philosophy inward to the subjective sphere and then arguing for a metaphysical distinction between mind and body, it has been widely thought that the focus on consciousness in itself leads to the separation of mind and body. From Locke, Hume, and Kant to Husserl in his transcendental phenomenology, it has seemed that the connections of mind and body have been cleanly severed. Yet a careful phenomenology of the experience of acting – of conscious volitional bodily *action* – leads instead to a subtle ontological intertwining of consciousness and body, and so of mind and the world in which it occurs.

Cogito ergo sum leads to *ambulo ergo sum*: inner awareness in embodied action leads to discrete awareness of one's own body and one's natural surroundings and to the connections between them. The phenomenology of action thus leads into an ontology of consciousness embedded in nature, in one's body, and – with futher empirical studies in neuroscience – in one's brain. We are beginning to turn our attention in this regard to the *substrate* of consciousness, to the natural, physical conditions on which our own consciousness depends.

Background Ideas

Our conscious experience is not only embedded in our bodily comportment in our natural environment; our experience is also embedded in our social environment. A close study of intentional content or meaning shows that our most familiar ideas – everyday concepts and rules of practice – presuppose very basic conceptual and practical structures that are extant in our surrounding culture. There is thus a deep dependence of our intentional contents on *background ideas* that virtually define the everyday world as we know it. Only by a sort of phenomenological-semantic archaeology, however, do we begin to appreciate this type of dependence. We may launch our study of this deep background of our intentional experience by starting with Husserl's conception of a "horizon" of background meaning and practice, Wittgenstein's notion of "ground propositions," and Searle's account of "background" capacities. However, we need to place these notions of background within a proper ontology of dependence. Here lies a crucial part of the *substrate* of consciousness – in the culture surrounding us, rather than in the neural processing within us. Indeed, background ideas have a life and status of their own, not in a Platonic or Fregean heaven of ideal meanings but in a realm of ideal meanings extant in our culture in the life world.

Intentionality Naturalized?

Contemporary philosophy-of-mind and cognitive science are largely wedded to a *naturalism* that assumes a functionalist physicalist ontology of mind. But functional-physical analyses of mind – of the physical inputs and outputs of different types of mental states – do not account for the crucial phenomenological features of consciousness: intentionality (and meaning), inner awareness, sensory qualia. What we need instead is a wider and more fundamental *ontology* that gives consciousness and nature their proper places in the structure of the world. A worthy start is Husserl's distinction between *formal* and *material* ontological categories.

In Husserl's ontology, for instance, the formal structure of states of affairs applies to entities in the material domains of nature (the physical), consciousness (the intentional), and culture (the social). We may thus begin to rethink the basic categorial structure of the world by considering fundamental categories of mind and world. We must distinguish different material types of properties of mental activity: those of consciousness (intentionality, etc.), those of nature (the physical causal conditions of consciousness), and those of culture (the social conditions of consciousness). But we must also distinguish different formal types of properties of mind: for instance, *intentional* relations of consciousness to its objects and *causal* relations of an event of consciousness to its causes and its effects. Without a fundamental ontology that draws such distinctions we cannot develop a *unified* account of mind and the world of nature, an appropriate phenomenological ontology.

Consciousness and Actuality

To understand the structure of the world in general, and the structure of consciousness in particular, we need to rethink our most familiar ontological concepts, which began with Plato and Aristotle on universals and particulars. A radically different type of ontology was envisioned by Whitehead, an ontology that would replace Aristotelian *substance* (centered on predication) with a fundamental type of *process* more attuned to twentieth-century physics. Today we might look to something like dynamic states in a relativistic quantum field (if we could understand such entities). Whitehead held that the most basic "actual entities" of the world are something like point events in a field of constant flux – out of which everyday objects emerge in great complexity. However, Whitehead distinguished what we may call *temporal* and *ontological* becoming. Whereas an "occasion" is formed by the process of temporal transition, any "entity" is formed by the process of becoming an entity, wherein an entity is ontologically dependent on a variety of other entities. This highly abstract form of *becoming* suggests a more fundamental kind of ontology, which may apply in instructive ways ultimately to the special case of consciousness.

Basic Categories

The doctrine of ontological *categories* began with Aristotle's list of ten. Husserl's distinction between formal and material categories ramified the very notion of ontological category, and of the categorial structure of the world, with instructive details applied to consciousness, nature, and culture. Whitehead's ontology of process, especially ontological becoming, suggests a deeper ontology of levels – or, as I prefer, "modes of being." We may begin to specify a more up-to-date *category scheme* by reflecting on these types of ontology. Three-facet ontology (distinguishing the form, appearance, and substrate of any entity) organizes three basic categorial structures, but there is more to the story. Marking various formal ontological distinctions, and organizing them in a structured system of categories, we may begin to frame a more systematic account of the order of things in general and of consciousness in particular.

Such is the story line that I mean to weave through and around the essays to follow.

Notes

1. That theory of intentionality is detailed in Smith and McIntyre 1982. A shorter version of the theory is presented by the same authors in 1989.
2. See Searle 1983.
3. Details on "indexical" content are found in Smith 1989. My account there extends and modifies traditional Husserlian phenomenology. Kindred spirits are at work in two books not directly linked with the phenomenological tradition: Searle 1983 (see the chapters on perception and action) and Perry 2001 (see Perry's account of "reflexive" content).

References

Perry, John. 2001. *Knowledge, Possibility and Consciousness.* Cambridge, Mass.: MIT Press.

Searle, John. 1983. *Intentionality.* Cambridge: Cambridge University Press.

Smith, David Woodruff. 1989. *The Circle of Acquaintance.* Dordrecht: Kluwer Academic Publishers.

Smith, David Woodruff, and Ronald McIntyre. 1982. *Husserl and Intentionality.* Dordrecht: D. Reidel.

1989. "Theory of Intentionality." In J. N. Mohanty and William McKenna, eds., *Husserl's Phenomenology: A Textbook*, pp. 147–79. Washington, D.C.: University Press of America.

The Picture

Many of us think visually, even when we conceptualize highly abstract phenomena. This is a phenomenological observation about the practice of phenomenology and ontology (for those like "us"). Indeed, I often draw pictures on the board while lecturing on the topics pursued in this book: structures of consciousness (intentionality, background, inner awareness, self-awareness) and structures of the world (ontological categories, the form of intentionality itself).

What follows, accordingly, is a pictorial organization of the structures of world and consciousness that are pursued in the essays to follow.

My students will recognize many of the elements of The Picture.

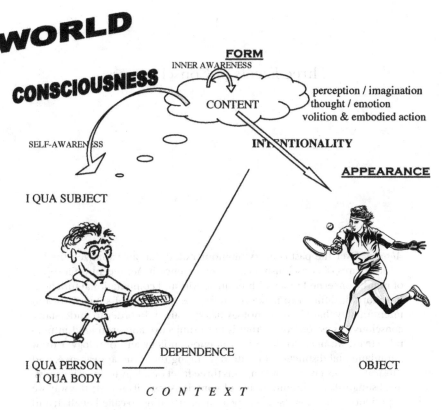

WORLD

CONSCIOUSNESS

FORM
INNER AWARENESS

CONTENT

perception / imagination
thought / emotion
volition & embodied action

SELF-AWARENESS

INTENTIONALITY

APPEARANCE

I QUA SUBJECT

DEPENDENCE

I QUA PERSON
I QUA BODY

OBJECT

C O N T E X T

BACKGROUND
Ideas ... Languages ... Practices

SUBSTRATE
NEURAL DYNAMICS ... BIOLOGICAL EVOLUTION ... QUANTUM FIELDS

CATEGORIES

FORMAL CATEGORIES
INDIVIDUAL   PROPERTY
STATE-OF-AFFAIRS

MATERIAL CATEGORIES
NATURE   CULTURE   CONSCIOUSNESS

BASIC CATEGORIES
Prehension Intentionality Modality Dependence Unity Plurality .....
ONTOLOGICAL BECOMING

1

Three Facets of Consciousness

Abstract: Over the past century phenomenology has ably analyzed the basic structures of consciousness as we experience it. Yet recent philosophy of mind, concerned more with brain activity and computational function, has found it difficult to make room for the structures of subjectivity and intentionality that phenomenology has appraised. In order to understand consciousness as something that is both subjective and grounded in neural activity, we need to delve into phenomenology and ontology. I draw a fundamental distinction in ontology among the form, appearance, and substrate of any entity. Applying this three-facet ontology to consciousness, we distinguish the intentionality of consciousness (its form); the way we experience consciousness (its appearance, including so-called qualia); and the physical, biological, and cultural basis of consciousness (its substrate). We can thus show how these very different aspects of consciousness fit together in a fundamental ontology. And we can thereby define the proper domains of phenomenology and other disciplines that contribute to our understanding of consciousness.

The Problem of Consciousness

Lately, philosophers and scientists have been looking for mind in all the wrong places. Physicalists of all stripes have focused primarily on the physical conditions of consciousness, from neural activity to computational

I am indebted to my colleagues in the PACIS project: Charles W. Dement, President of Ontek; Stephen DeWitt, John Stanley, and Anthony Sarris, all of Ontek; and Peter M. Simons of the University of Leeds. Thanks to Tony Sarris for the ISO reference. Thanks further to Chuck Dement for numerous discussions of systematic formal ontology. I bear responsibility, nonetheless, for what is made of the three-facet distinction in the present chapter.

function.[1] Meanwhile, humanists – historicists, postmodernists, culture critics – have looked primarily to the cultural conditions of our discourse, as if consciousness did not exist in its own right (expressed in art and literature) but is "theorized" in a cultural tradition of phenomenology or science or humanistic discourse. Obviously, we have much to learn from the empirical sciences about boson, atom, organism, evolution, and brain – and from humanistic observations in art, literature, and cultural history and criticism. But this learning is informed by further disciplines that are not "empirical" or "naturalistic" or indeed "humanistic" in the received ways. If we are to understand the mind, we must understand more clearly the philosophical disciplines of phenomenology and ontology, because these disciplines define the place of mind in a world further detailed by the scientific disciplines of neuroscience, evolutionary biology, and quantum physics, as well as the humanistic disciplines of literary, artistic, and cultural criticism.

Let us begin with a fundamental principle of ontology. The nature of any entity, I propose, divides into three aspects or *facets*, which we may call its form, appearance, and substrate. In an act of consciousness, accordingly, we must distinguish three fundamentally different aspects: its form or intentional structure, its appearance or subjective "feel," and its substrate or origin. In terms of this three-facet distinction, we can define the place of consciousness in the world. The aim of this chapter is to lay out this distinction in the nature of consciousness, and to draw out its implications for phenomenology and ontology, as distinct from purely naturalistic philosophy of mind. (I do not focus here on humanistic theory, although I think the morals to follow have relevance for humanistic as well as naturalistic theory of mind.)

Consciousness is the central concern of phenomenology. Although there is more to mind than what we consciously experience, our theory of mind must begin with the salient part of mind, conscious intentional experience. Consciousness is characteristically a consciousness "of" something, as Husserl stressed circa 1900, and this property of directedness he dubbed *intentionality*. The literature of phenomenology – in Husserl, Heidegger, Sartre, Merleau-Ponty, Ingarden, Føllesdal, and others, with roots in Kant, Hume, Descartes, and still earlier thinkers – has analyzed a rich variety of structures of intentionality in perception, imagination, thought, language, and action, along with properties of subjectivity, intersubjectivity, temporality, and the unity of the subject or self. For the discipline of phenomenology, there is no problem about the nature or existence of consciousness: we experience it firsthand throughout our waking life, and we have ways of studying it carefully.

For recent philosophy of mind, however, consciousness has seemed problematic, either in its nature or in its very existence, because it seems to escape the story told by the physical sciences. "Consciousness is what makes the mind-body problem really intractable," Thomas Nagel observed, rightly, wryly, and presciently in 1974 (Nagel 1974). As cognitive science developed over the next two decades, moving from artificial intelligence into neuroscience, consciousness regained center stage. The function of mind in mediating behavior, in problem-solving computation, in evolutionary adaptation, and the like did not seem to involve the subjective qualities of sensation, dubbed qualia, or the felt character of consciousness as directed toward objects in the world around one. Nonetheless, by 1990 neuroscientists were measuring properties of neural activity (such as spiking frequency) associated with consciousness, and so consciousness became a respectable phenomenon of scientific investigation. "Consciousness studies" emerged with large interdisciplinary conferences in Tucson in the 1990s. Still, amid the excitement even in popular media, David Chalmers (1996) echoed Nagel's sentiment in declaring consciousness the "hard" problem for our theory of mind. Chalmers struck a nerve.

Yet is it not odd to find consciousness problematic? What if someone declared that we do not know what language is, or that its existence is uncertain? We all speak a language such as English or Japanese. Grammarians have charted its basic forms such as the verb or noun phrase, and linguists have analyzed its "deep" structure. How the brain functions in the production and understanding of language is a further matter of empirical neuroscience; how speech and writing emerged in our species is a matter of evolutionary biology; how our language shapes our society and politics is a matter of social-cultural theory. But the syntax and meaning of modern English are familiar, more or less, to its speakers. Similarly, the shape and meaning of our everyday experiences of perception, thought, and action are familiar to us all, more or less. These forms of consciousness have been studied by phenomenologists, much as linguists have studied forms of language. How the brain functions in consciousness, how our forms of experience evolved in the species *Homo sapiens sapiens,* how our consciousness is shaped by our language, culture, and politics – these are further matters. But how can consciousness itself be thought problematic or its basic forms obscure?

There is a widespread opinion that science alone will explain the work-ings of the world, including our own minds and thus consciousness. This idea goes under the positive banner of "naturalism" or meets the pejo-rative charge of "scientism." This attitude is expressed with character-istic verve, in his recent book *Consilience* (1998), by biologist Edward O. Wilson, famous for his studies of ants and for his conception of socio-biology. Let us quote at length (the only way to evidence "attitude," albeit in the way of humanists):

Belief in the intrinsic unity of knowledge ... rides ultimately on the hypothesis that every mental process has a physical grounding and is consistent with the natural sciences. The mind is supremely important to the consilience program [of unity] for a reason both elementary and disturbingly profound: Everything that we know and can ever know about existence is created there.

The loftier forms of such reflection and belief may seem at first to be the proper domain of philosophy, not science. But history shows that logic launched from introspection alone lacks thrust, can travel only so far, and usually heads in the wrong direction. Much of the history of modern philosophy, from Descartes and Kant forward, consists of failed models of the brain. The shortcoming is not the fault of the philosophers, who have doggedly pushed their methods to the limit, but a straightforward consequence of the biological evolution of the brain. All that has been learned empirically about evolution in general and mental process in particular suggests that the brain is a machine assembled not to understand itself, but to survive. Because these two ends are basically different, the mind unaided by factual knowledge from science sees the world only in little pieces. It throws a spotlight on those portions of the world it must know in order to live to the next day, and surrenders the rest to darkness. (Wilson 1998, 96)

What we have here is failure to communicate, between philosophers and scientists. (1) It was philosophers – Descartes, Kant, and Husserl – who taught us the principle, "Everything that we know and can ever know about existence is created there [in the mind]." (2) The history of mod-ern philosophy includes much more than failed models of the brain; Descartes and Husserl developed successful models of consciousness, of mind as experienced, precisely what is now found "hard" for empirical neuroscience. (3) Although the brain did not evolve to understand it-self, in humans it seems to be on the verge of producing, Wilson thinks, a scientific theory of its own physical and evolutionary function – and, I think, a philosophical theory of consciousness. (4) Most of the great modern philosophers – notably Descartes, Kant, and Husserl – theo-rized in the face of factual knowledge from science *cum* mathematics in their day; they also appreciated, however, the importance of intro-spection when attending to the mind. (5) It is a hallmark of modern philosophy – and ultimately philosophy of science – to delimit knowledge of empirical fact and that of logic and mathematics, and thus to define the limits of both a posteriori and a priori knowledge; today in philosophy-and-science of mind we need to understand the boundaries and inter-relations between the more empirical and the more "formal" aspects of consciousness.

A different view, from the formal side of natural science, is proposed by mathematical physicist Roger Penrose in *Shadows of the Mind* (1994). From Kurt Gödel's incompleteness theorem in mathematical logic, Penrose argues that consciousness cannot be a process of computation in the technical sense originally defined by Alan Turing; then, from consider-ations of quantum mechanics and the microstructure of neurons in the human brain, Penrose argues that we need a noncomputational quan-tum physics to explain how consciousness can arise in neuronal activity. I cannot evaluate the controversial speculations in Penrose's book, but if he is right then consciousness is defined by a very different kind of "formal" mathematical structure than anything philosophers of mind have been considering previously. What I like in Penrose's vision is this type of abstraction. The mathematical form of a piece of physical theory is integral to its content, and mathematical form is suggestive of ontological form. The subtitle of the Penrose book is *A Search for the Missing Science of Consciousness*. When we have finished the "science" of consciousness, its physics and its evolutionary biology, there will still be something missing in our account of consciousness. What is missing in all current "natural-istic" thinking about consciousness is the relevant phenomenology and ontology, and their integration.

The "loftier forms" of naturalism are what attract the philosopher. I believe in the unity of knowledge. I believe, moreover, in the unity of the world: one world in which physical, mental, and cultural phenomena take their interweaving places. And I believe that every mental process has a physical grounding and is consistent with the natural sciences. (In fact, I am quite partial to the metaphor of "ground" in ontology, as we shall see.) So far, naturalism: both methodological and ontological (these need to be distinguished).

However, the structure of intentionality – call it "formal" or "transcen-dental" or something else – does not flow easily from empirical, "natu-ralistic" studies of the brain or bodily behavior or physical system alone. The "logic" of intentionality in phenomenology, methodically launched from introspection alone, has a powerful thrust and carries us far (con-trary to what Wilson claims in the preceding quotation). However, I must concur, the theory of intentionality carries us in different directions than empirical science: into structures of consciousness in phenomenology, and indeed into structures of thought and inference in logic and semantics (concerning how we reason and represent things in thought and language). "Formal" ontology, too, moves in different directions, positing fundamental categories of existence such as Individual, Property, Relation, Number, Part, and so forth. Both phenomenology and ontology are crucial to a unified system of knowledge – of a unified world. And both carry us beyond naturalism: their results should be consistent with natural science, but the proper results of phenomenology and ontology are not simply amassed in empirical investigation in the natural sciences alone.

When we want to see the world as a whole and not in the "little pieces" so effectively modeled by physics, chemistry, and biology – when we want to see the *unity* of the world – we must inform natural science with fundamental ontology. Much as physics needs mathematics to structure its empirical content, so natural science in general needs ontology – or metaphysics – to structure empirical content. And when we turn to the nature of mind itself, the empirical analysis of our own consciousness is pursued expressly and methodically by phenomenology. Moreover, it is ontology that must define the *type* of relation that holds between mind and its grounding in brain activity. This is a matter of formal ontology rather than of empirical investigation per se.

Wilfrid Sellars (1963) contrasted two ranges of theory that define respectively the "manifest" image and the "scientific" image of man, that is, ourselves and our world as understood by common sense and as described by modern science. Similarly, in *The Crisis of European Sciences and Transcendental Phenomenology* (1970b/1935–38) Husserl distinguished the "lifeworld" from the "natural world," that is, the world as we experience it in everyday life and as we "mathematize" it in physics. Mathematical physics (in all its well-earned glory) is an *abstraction*, Husserl held, from the world as experienced in everyday life. Consequently, Husserl said, we must confront "the paradox of human subjectivity": how can I be both subject and object of consciousness, both a conscious subject and an object in nature? Husserl did not clearly foresee the "mathematization" of thought in the computer model of mind. Yet today's controversy about mind as computer (whatever the architecture, classical or connectionist) is but the application of mathematical modeling or "mathematizing" to mind as opposed to physical activity like planetary motion. Thus, Husserl's "paradox" foreshadowed what today is the "hard" problem of consciousness: how can consciousness be both a *subjective* character of experience and an *objective* property of the brain – a computational structure implemented in neural networks evolved over the natural history of the human species on the planet Earth in the cosmos that took shape since the Big Bang over 12 billion years ago amid fields of gravity, electromagnetism, and quantum superposition?

Husserl distinguished phenomenology from both everyday knowledge and scientific knowledge, and he distinguished "formal" ontology from "material" ontologies of Body, Culture, and Consciousness (as the distinction is reconstructed in D. W. Smith 1995). The point to stress here is that the world is characterized in different parts and levels in these different ranges of theory, and the philosophy of mind must respect these differences of theory.

Only by understanding more clearly both phenomenology and ontology, along with the natural sciences (as well as the humanities), can we understand the place of consciousness in the world. That is the loftier moral of this essay. The specifics to follow concern the ontology of the three aspects or "facets" of consciousness, and the role of phenomenology in such an ontology.

Phenomenology and Ontology

Ontology (or metaphysics) is the science of being: as Aristotle put it, being *as* being. Where the special sciences – physics, chemistry, biology, psychology, and the like – are sciences of particular kinds of beings, ontology is the general science of what it is to be a being (and perhaps of what it is to be).

Phenomenology is the science of consciousness: as Husserl put it, of consciousness *as* we experience it. Phenomenology begins in the description of conscious experience from our own point of view as subjects or agents: "I feel angry," "I see that volcano," "I think that Plato was ironic," "I will [to act so that I] stroke this tennis ball cross-court," and so on. The intentionality of consciousness is evident in our own experience: I am conscious "of" or "about" such and such.

Now, ontology and phenomenology interact in our overall theory of consciousness and its place in the world. For our experience – in emotion, perception, thought, and action – is informed by our understanding of the world around us, by our ontology, implicit or explicit. And as we practice phenomenology, we use our ontology, implicitly or explicitly, in order to describe our experience, its intentional relation to objects in the world, and the things we are conscious of in perception, thought, and action. In this way, phenomenology is ontological. But ontology itself is phenomenological insofar as it recognizes the existence of our own consciousness, as we must in saying what exists.

It may be surprising to speak of ontology within the practice of phenomenology. For did not Husserl, in *Ideas* (books I and II, 1969/1913 and 1991/1912ff.), enjoin us to bracket the existence of the surrounding world of nature (and culture) in order to describe the structure of our consciousness? Here lies confusion. Husserl assumed a good deal of "formal" ontology – concerning individual and essence, part and whole, dependence, and so on – precisely as he sought to describe the essence of intentionality in phenomenology; and bracketing the region of nature (and the region of culture) leaves the region of consciousness, with the "material" ontology of consciousness as part of phenomenology (see D. W. Smith 1995). Heidegger followed suit, in *The Basic Problems of Phenomenology* (1988/1975/1927), assuming his own formal categories in describing structures of our existence and comportment; indeed, Heidegger insisted that phenomenology *is* "fundamental ontology," and so fundamental ontology is essentially phenomenological. Philosophy today has lost sight of the intimate connection between our saying what there is and our saying how we experience what is.

Let us approach the nature of consciousness and its place in the world by laying out a very basic ontological distinction, a distinction we rarely make explicit but assume deep in the background of a good deal of our theorizing about the world.

Three-Facet Ontology

Everything in the world – every entity whatsoever – has a nature that divides fundamentally into three aspects we shall call *facets*: its form, its appearance, and its substrate. Thus:

1. The *form* of an entity is how or what it is: its whatness or quiddity – the kinds, properties, relations that make it what it is.
2. The *appearance* of an entity is how it is known or apprehended: how it looks if perceptible (its appearance in the everyday sense), but also how it is conceived if conceivable, how it is used if utilizable – how it is experienced or "intended" as thus and so.
3. The *substrate* of a thing is how it is founded or originated: how it comes to be, where it comes from, its history or genetic origin if

temporal, its composition or material origin if material, its phylogenetic origin if biological, its cultural origin if a cultural artifact –
in short, its ecological origin in a wide sense, and ultimately its
ontological origin in basic categories or modes of being.

The three *facets* of an entity (in this technical sense) are categorially
distinct aspects of the entity, with important relations among them, as we
shall be exploring. This distinction of aspects we may call the *three-facet*
distinction (Figure 1.1).

Distinctions among form, appearance, and foundation or origin have
been drawn in philosophy since its inception. Plato distinguished concrete things from their forms, and appearance from reality, and posited
forms as the foundation of being. Before Plato, Anaximander assessed
the material composition of things and envisioned their origin or foundation in something more basic (an archaic quantum field?); he even
foresaw biological evolution, 2,500 years before Darwin. In more recent
centuries, epistemologists from Descartes to Kant distinguished things
from the ways they are known, while idealists like Berkeley put mind at
the foundation of reality and materialists reduced mind to matter. What I
am proposing, however, is to unify the distinctions among form, appearance, and substrate, and then to elevate the three-facet distinction itself
to an axiom of fundamental ontology – and so to structure ontology itself
(in one way) along these lines.

The structure < Form, Appearance, Substrate > thus defines a special
system of ontological categories. For the world is structured importantly,
at fundamental joints, by this three-facet distinction. The distinction presupposes that the world includes attributes (of entities), minds (to which
entities may appear), and contexts of foundation or origin (from or within
which entities may come to be). There may be possible worlds that lack
such things, but our world has this much structure, and our ontology and
phenomenology are accountable to this three-facet structure of the world.
These three categories do not form a sequence of mutually exclusive and
collectively exhaustive *summa genera* of entities, as do the Aristotelian categories (roughly, Substance or Individual, Species, Quality, Quantity, etc).

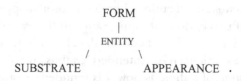

FIGURE 1.1. The three facets of an entity.

Rather, the categories of Form, Appearance, and Substrate order or rank three fundamental ways an entity in our world is defined: by relation to its form, to its being known or "intended," and to its ground or origin. If you think about it, these categories define three fundamentally important and importantly different areas within the *nature* of any entity (in a world such as our own). Thus, the entity itself is *distributed* in its being through these three aspects of form, appearance, and substrate: that is its nature or essence.

There are other fundamental divisions in the structure of the world. But the division < Form, Appearance, Substrate > marks one crucial ordering in the nature of things. To appreciate its significance, we shall work through some examples.

Importantly, the division among form, appearance, and substrate is a division of *structure* in the nature of an entity rather than a division among three intrinsically distinct types of property. In principle, the same thing might be part of the appearance, form, and substrate of an entity. The green of a leaf – say, of a California Live Oak tree – is part of its appearance to the human eye, part of its intrinsic form, and part of its evolutionary history (in the role of chlorophyll). Thus, the property green plays three different *roles* in the form, appearance, and substrate of the leaf, and these three facets are themselves defined by the roles played. That is, Form, Appearance, and Substrate are defined by *roles* played in the nature or essence of an entity.

An instructive parallel to the three-facet distinction can be drawn in biology: in systematics, the science of diversity among living things (Mayr and Ashlock 1991). Thus, biologists today *define* a species by principled reference to its character, its observed specimens, and its evolutionary descent. (Exactly how a species is defined in these terms has been vigorously debated; I abstract the terms of debate only.) Imposing our terminology: the form of a species is its genotype, its appearance (to the scientific community) is its phenotype or observable traits (starting with a definitive specimen called its holotype), and its substrate or origin consists in its path of phylogenetic descent from ancestor species. These three aspects of a species are given canonical places in defining the species in modern biological theory, and we may see in this empirical theory something more like a "formal" division in the nature of all things, not just of evolving species of living things. (N.B. In biological systematics, a "category" is defined not as a high-level grouping or *summum genus*, as Aristotle originally used the term, but rather as a rank of taxa or groupings. In the long run I too would define the term "category" in a more special way, but for present purposes let us mean by the term simply an important group or classification of things.)

Insofar as biological systematics provides one model of three-facet ontology, we are abstracting or factoring out from the empirical theory of species the formal structure of three facets, which we would apply to any kind of entity at all (in a world such as our own). This kind of abstracting is the proper work of formal ontology, and the three-facet distinction is a formal ontological distinction, applying by hypothesis to any kind of entity whatsoever. The significance of the three-facet distinction lies in the different ways in which something can be defined in its being, in its fundamental nature, by its form, appearance, and substrate.

Husserl is one systematic philosopher who recognized what we are calling the three facets of an object. The form of an object Husserl called its "essence" (*Wesen*, from *Was-sein*). The sensible qualities of a material object Husserl called its "appearance" (*Erscheinung*), or more generally its "way of being given" to consciousness, which aligns with the "object as intended." And the substrate of an object encompasses what Husserl called its "horizon," its *Umwelt* (surrounding world), and indeed its ontological "foundation," that on which it is "founded" (by *Fundierung*). What is unusual in Husserl's philosophy is the principle that the *essence* of any object includes the ways in which it can be known or intended, the ways it is "constituted" in consciousness. It was Husserl who first explicitly defined "formal" ontology, as specifying categories ("formal essences") that apply to different "regions" ("material essences") such as Nature, Culture, and Consciousness. (The details are drawn in D. W. Smith 1995.) The three-facet distinction belongs to formal ontology in this sense. However, Husserl himself did not join the three facets into the canonical division I am proposing.

The three-facet distinction, then, is a higher-order formal structure that orders the nature or essence of an entity. Because this structure applies to consciousness itself, we can use the three-facet distinction to look at the ontology of consciousness. But first we must address the distinction generally.

The Three Facets of Diverse Entities

To see how the three-facet distinction works, and to begin to appreciate its scope, let us apply it to some very different kinds of entities.

Consider this piece of quartz found in my garden. Its form includes its shape, its color, and its type, quartz. Its appearance includes what it looks like from various angles and under various lighting. And its substrate includes its physical crystalline structure, as well as its geological genesis from great heat and pressure in the crust of the planet Earth.

Consider now an electron. Its form includes its mass, charge, and spin. Its appearance includes its observable position and momentum, its electron-microscope image, and so on. And its substrate includes the matter field (from which it emerges per quantum field theory). So even a basic physical particle has its three facets.

Now consider this pencil. The form of the pencil is its structure of graphite in wood plus its function in writing and drawing. Its appearance includes what it looks like and what it feels like in my hand in writing. The substrate of the pencil is its origin. It is made of certain materials, including wood, graphite, paint, tin, rubber. Each material has its physical-chemical structure. Moreover, these materials are produced only in specific parts of the world, in specific cultures, their trade following established routes. Furthermore, the substrate of the pencil includes the historical development of writing, writing instruments, and the invention of the pencil. So the pencil's substrate includes not only its physical composition (down to quantum structure) but also its cultural genesis.

Next consider the tool of the century: the computer. The International Organization for Standardization has defined what is called the ISO three-schema architecture for database design, distinguishing a computer program ("conceptual schema"), its implementation in hardware (plus operating system, etc.) ("internal schema"), and the user's presentation of what the program does ("external schema"). These three aspects of a computer system are precisely what we are calling its three facets: its form, the program; its substrate, the hardware; and its appearance, the user interface. This familiar distinction, we now begin to see, reflects a deep ontological distinction in the nature of things far beyond computers.

Finally, consider a human being, an individual such as Napoleon. His appearance is well known: his facial structure, his small stature, his posture with hand in vest. His form is his individual character as a person, an intentional subject living in a culture in the natural world, his body having various traits. And his substrate is what makes this individual possible: his genetic heritage, his birth on Corsica, the French Revolution, and the army in which he developed his power – as well as the wider physical, biological, and cultural conditions of humanity.

Observe how naturally the three-facet distinction applies to such diverse kinds of entities. The concepts of form and appearance are relatively

familiar; the concept of substrate is not. Indeed, notice how wildly different are the things that serve as *substrate* for different entities: materials
or parts from which an object is composed; the field in which a physical
particle exists; the genesis of an individual through time; the evolutionary
track (or "clade") of a biological species; the cultural history and use of
an artifact; the hardware that implements a computer program; the life
trajectory of an individual human being; even the cultural genealogy of
our values (in Nietzsche's idiom) and of our language games and other
forms of life (in Wittgenstein's idiom). What these things share, what
makes these things play the role of substrate in very different entities, is
the form of ontological *derivation* or *emergence* (in different ways!) from
things more fundamental, the form of ontological *foundation* or *dependence* on things in the wider context of the entity. Again, the three-facet
distinction belongs to "formal" ontology.

Now let us apply the three-facet distinction to – of all things – consciousness itself.

The Three Facets of Consciousness

An act of consciousness – my experience of thinking, seeing, or doing
such and such – is an entity with three facets:

1. Its *form* is its structure of intentionality, its being directed from
   subject toward object through a content or meaning, with inner
   awareness of itself ("apperception").
2. Its *appearance* is how I experience it, "what it is like" for me to live
   or perform this act of consciousness.
3. Its *substrate* is its origin or background in conditions including brain
   activity, psychological motivation, cultural ideas or practices, and
   the biological evolution of this form of mind.

The three facets of an act of consciousness are mapped in a diagram in
Figure 1.2.

According to this three-facet ontology, an act of consciousness is *distributed* in its nature through its form, appearance, and substrate. This is
not to say there are three kinds of entities bound together, say, items of
brain, meaning, and "feeling." Rather, a particular mental act is one entity
with a nature that divides into three fundamentally different aspects or
facets. There are systematic relations, including dependencies, among

STRUCTURE OF INTENTIONALITY
SUBJECT—ACT—CONTENT—>OBJECT
INNER AWARENESS

FORM
|
ACT OF CONSCIOUSNESS
/ \
SUBSTRATE   APPEARANCE

NEURAL-PHYSICAL-BIOLOGICAL CONDITIONS   PHENOMENOLOGICAL CHARACTERS OF
PSYCHOLOGICAL CONDITIONS   PERCEPTION, THOUGHT,
CULTURAL CONDITIONS   EMOTION, VOLITION ...

FIGURE 1.2. The three facets of an act of consciousness.

these facets, but that is a further story. First we must appreciate the fundamentally different roles these facets play in defining consciousness. As we bring out these differences in facets, we can carve out the role of phenomenology in understanding mind and its place in the world.

The Structure of Intentionality and Inner Awareness

I am assuming a basic theory of intentionality (elaborated in Smith and McIntyre 1982). This account of intentionality draws on a long history, but the main ideas were synthesized adequately first by Husserl. In recent work in philosophy of mind and cognitive science, Searle (1983) comes closest to this model of intentionality.

Consciousness occurs in concrete states or events called "acts" of consciousness: when I see that bird overhead, when I think that Plato was ironic, when I feel angry about the president's speech, when I run with the volition to catch a bus, and so on. Such an act of consciousness is *intentional*, or directed toward something, called its object (the bird, Plato, the speech, my catching the bus). As we say, it is "of" or "about" that object. The object is prescribed by the content of the act. And the act is experienced or performed by a person, called its subject. This structure is analyzed by laying out cases and marking distinctions among subject, act, content, and object (for detail, see Smith and McIntyre 1982). The point to consider here is where this structure plays in the ontology of the act of consciousness, in the three-faceted nature of the act.

The fundamental structure of intentionality, we assume, is this:

$$\text{subject — act — content} \longrightarrow \text{object.}$$

The act is distinct from the object (unless the act is self-referential). The subject is distinct from the act, and from the stream of consciousness in which the act occurs as a transitory part. The content is an idea, image, concept, thought (proposition), volition, and so on: a "meaning" that Husserl called noema, updating the ancient Greek term for "what is known." Importantly, the same object can be "intended" through different contents in different acts.

In practicing phenomenology, when I reflect on an act of consciousness as experienced, I "intend" the act in a second act of reflection focused on the first act. In this reflection the first act "appears" to the second. Classical phenomenology was much exercised about the best methodology for reflecting on our experience. However we do it, let us assume, I carry out an act of reflection – or introspection – on my own conscious experience. And in phenomenological reflection the intentionality of the given act is part of its "appearance" in the reflective act. So the given act's intentionality is part of its *form* but also part of its *appearance* in reflection.

Furthermore, we may directly experience the intentionality of an act of consciousness, without retreating into a reflection on it. For when I am conscious of something, say in perception, I have a prereflective *inner awareness* of this consciousness-of-something. On this neoclassical view, consciousness is consciousness-of-something and *eo ipso* consciousness-of-itself. In inner awareness, then, intentionality "appears" to me in having a conscious experience of such and such. So intentionality is part of the *appearance* of the act already in inner awareness. (This form of inner awareness is analyzed in D. W. Smith 1989.)

By contrast, we do not directly experience (in inner awareness) the *substrate* of an act of consciousness, notably its grounding in brain process and cultural history. Nor does the substrate of an experience submit to phenomenological analysis in reflection or introspection. In modern times we all have some knowledge of the fact, empirically discovered and pursued in neuroscience, that what we are thinking, perceiving, or dreaming depends on what is happening in our brains. And in postmodern times we all have some appreciation of the fact, frequently observed in the wake of Marx, Nietzsche, and Freud, that what we think and value depends on long-standing cultural conceptions, assumptions, linguistic practices, and political institutions. But these background conditions must be distinguished from the intentional structure of consciousness itself, thus separating the form and appearance of consciousness from its substrate. (Compare D. W. Smith 1999b on "background ideas" in the cultural substrate of intentionality.)

The Ontology of Form and Appearance

Given the preceding model of intentionality, we can say more about the ontological status of the form and appearance of an entity.

The *form* of an entity, we said, consists of its kinds, properties, and relations. These are "universals" in the traditional sense. I shall not here address the full range of issues about the existence of universals (ably and succinctly assayed by Armstrong 1989, 1997), but a couple of points stand out in present discussion. First, some universals depend for their existence on intentional acts of consciousness and associated cultural practices, although most do not. The property of being a fork, for instance, could not exist unless people had developed the tradition of eating with a utensil of that shape. (See Thomasson 1999, on similar issues of dependence.) Moreover, if I am using a fork to pry open a box, it is not in that context bearing the property of being a fork. Second, a universal is distinct from any concept that represents it. The property of being an electron does not depend for its existence on anyone's having a concept of it; when someone thinks about an electron, the concept "electron" is part of the content of the act of thinking but is distinct from the property of being an electron. Universals rather than any associated concepts make up the form of an entity. Third, a universal is distinct from its *instance* in a particular entity. Aristotle called such instances "accidents"; Husserl called them "moments"; recent philosophers (following Donald Williams's usage) call them "tropes." Strictly speaking, the form of an entity – in the three-facet distinction – is realized in a complex comprising moments or tropes that are instances of universals. For simplicity, however, in this essay I shall simply speak of an entity's form and its constituent kinds, properties, and relations. (Still, the distinction between moments and universals, or "ideal" essences, does important work in a Husserlian philosophy of mind *pace* D. W. Smith 1995.)

The *appearance* of an entity, we said, consists of how it is known or, we may now say, "intended" in appropriate acts of consciousness. This talk of "how" is ambiguous between the properties that appear or are intended and the contents through which they appear. (Husserl carefully distinguished these: see *Ideas* I, 1913, §42.) When I see that green leaf, for instance, the content "green" in my visual experience is one thing, and the color itself in the leaf is another thing. Science tells us that the color green is dependent not only on the wavelength of the light reflected from the surface of the leaf but also on the interaction with the observer. Nonetheless, the color in the leaf is distinct from the sensory-conceptual content in my experience. Moreover, as Husserl noted, the same color will "look" different under different lighting conditions. The properties in the appearance of the leaf, in its three-facet nature, are distinct from the concepts or sensuous qualities (so-called qualia) that are part of my visual experience that intends the leaf and presents it as green.

Clearly, appearance is to be studied in different philosophical disciplines, in phenomenology (addressing its role in intentional consciousness) and in ontology (distinguishing an appearing color itself from the intentional contents that present it).

Bearing in mind these amplifications of form and appearance in general, we turn to phenomenology as the study of the form and appearance of consciousness.

"Transcendental" Phenomenology and the Study of Consciousness

Consciousness has seemed difficult to study in a disciplined way because it is hard to separate in a disciplined way the different features of consciousness. The distinction between three facets of consciousness helps to define the domains of different disciplines that study consciousness. Purely descriptive phenomenology describes the *appearance* of consciousness in our own experience: the character of consciousness as we experience it in different types of experience. Analytic ("eidetic") phenomenology analyzes the *form* of consciousness: the formal structures of intentionality (already noted in description of our experience). Empirical sciences investigate the substrate of consciousness: the conditions in which it arises in different forms (noted, roughly, in phenomenological analysis). Neuroscience develops the theory of how neural activity gives rise to consciousness in different forms, taking these forms somewhat for granted. Evolutionary biology develops the theory of how different life forms evolved, including animals (and plants?) with the capabilities of consciousness, which begins in sentience and response in low-level organisms (perhaps even in DNA structures themselves if we are to believe some abstractions about the transfer of "information" in very different levels of physical reality). Cultural history develops the theory of how particular forms of consciousness evolved in human history, including the types of conscious and often collective thinking that we call storytelling, philosophy, mathematics, and empirical science.

Ontology – formal ontology – distinguishes not only the three facets of any entity but the fundamental structures of the world, and so ultimately the place of consciousness in the world, including its relation to brain, species, culture. In this way, ontology defines the parameters of the different disciplines that study consciousness in very different ways.

In particular, the three-facet distinction in ontology provides a way of defining phenomenology that preserves its original insights without contravening the wonders of today's neuroscience, not to mention evolutionary biology, cosmology, and cultural history. There is no denying the relevance of brain, biology, physics, and culture to consciousness. Yet phenomenology is a different discipline than the empirical natural sciences and the hermeneutic "human sciences." Husserl labored hard to distinguish phenomenology as a "rigorous science" that is distinct from both the natural and cultural sciences. He called this discipline "transcendental."

In practice Husserl (1912–13) defined "pure" or "transcendental" phenomenology as the study of consciousness while bracketing the surrounding world of nature and culture. This methodology Husserl allied with an ontology of distinct essences or "regions" called Consciousness, Nature, and Spirit or Culture. To these "material" essences Husserl applied the "formal" categories of Individual, Moment or Instance-of-Essence, and State of Affairs. Phenomenology studies intentionality, the central feature in the essence of an act of consciousness (formally an individual). But what is "transcendental" about this phenomenology and attendant ontology?

Dagfinn Føllesdal (1969/1982) focuses phenomenology on the "meaning" things have for us in intentional experience. Husserl called this meaning or content the noema of an act of consciousness. "Transcendental" phenomenology then focuses on *meaning*. Meaning is, on this view, an objective content of intentional experience that can be shared by other acts and, in some ways, expressed in language or in pictures or other media of expression and communication.

J. N. Mohanty (1985) has expounded transcendental philosophy and phenomenology, looking to Kant, from whom Husserl appropriated the term "transcendental." In the spirit of Kant, phenomenology would study the basic conceptual and sensory structures of our experience. More precisely, transcendental phenomenology is "reflection upon consciousness in its object-constituting role" – that is, intentionality, beginning by "delineating the structure of a noema" (p. xix). What is "transcendental" are the noematic meaning structures through which consciousness intends objects. However, we can no longer hold with Kant that such structures are necessary and a priori (p. xxix).

Robert Sokolowski (2000) traces the definition of "transcendental" from Kant back to the medieval philosophers. For the Scholastics, "transcendental" categories apply not to concepts but to beings: the categories of Unity, Being, and the like are "transcendental" because they pertain to absolutely everything. Kant revised this notion, with his Copernican revolution, so that what is "transcendental" are our fundamental concepts: the categories of the understanding, through which alone we can conceive of objects in the world. (Kant explicitly revised the Aristotelian-Scholastic categories by pulling them up from objects into concepts.) Phenomenology then transcends the Scholastic and Kantian conceptions of "transcendental." As Sokolowski shrewdly observes, "Transcendental phenomenology is the mind's self-discovery in the presence of intelligible objects." Whence: "to think about intentionality in all its forms . . . is transcendental phenomenology."

I concur that transcendental phenomenology is the study of intentionality. But I propose to rethink the traditional conceptions of "transcendental." We may say the three-facet distinction in ontology is itself "transcendental," radically updating the medieval notion, insofar as this distinction defines a *fundamental formal ontological structure* that applies to any entity in our kind of world (from a biological species to a computer system). Indeed, in terms of the three-facet architecture we can redefine the role of phenomenology in studying the nature of mind.

The appearance of an act of consciousness is studied in what Husserl called "phenomenological description," description of experience as lived. The form of an act of consciousness – the details of its intentional structure – is studied in what Husserl called "eidetic analysis" of the act, abstracting or factoring out its form or *eidos*, namely, its being directed from subject to object via content. These are the two stages of method Husserl (1913) used in the practice of "transcendental phenomenology." Now, the substrate of an act of consciousness is studied not by phenomenology, but by the natural sciences (in neuroscience, evolutionary biology, physics) and the cultural sciences (in criticism and history). It is precisely these aspects of substrate that are bracketed in practicing phenomenology.

What is "transcendental" in phenomenology, then, is its focus on the form and appearance of consciousness, as distinguished from its substrate. But what makes this focus "transcendental" is its marking the formal-ontological distinction among form, appearance, and substrate. For what is "transcendental" in philosophy, in the present scheme, is "formal" ontology: seeking very basic forms of being. We need not cling to the term "transcendental"; what is important in the study of consciousness is observing the differences among form, appearance, and substrate. If we observe these differences, we should not rule out the relevance of neuroscience and biology for the phenomenological and ontological study of consciousness: empirical studies of brain activity may reinforce or even help to clarify phenomenological observations by showing where in the brain a particular process or quality of consciousness is produced. Still, the analysis of form and appearance in consciousness is the business of phenomenology, informed by ontology. (Compare D. W. Smith 2000b and 2000c.)

"Naturalistic" Philosophy of Mind and the Study of Consciousness

The lessons of classical phenomenology are being rediscovered in recent philosophy of mind *cum* cognitive science, which begins with the metaphysics of naturalism, holding roughly that everything, including mind, is a part of nature and so – turning to epistemology – is to be studied following the methods of the natural sciences. Within this context we need to draw out the implications of the three-facet distinction in formal ontology.

As noted, it is the discipline of neuroscience, not phenomenology, that must teach us about the inner workings of the brain, how it produces consciousness, how it implements the structure of intentionality – the form of consciousness – in human beings and other terrestrials. When neuroscience-minded philosophers like Patricia Churchland (1986) and Paul Churchland (1995) eliminate the propositional attitudes of belief and the like in favor of neuronal activity, or collapse consciousness into neural flashes, they have limited their view of consciousness to its neural substrate. But there is more in view: there is form and appearance, where intentionality and its subjectivity reside.

Casting a wider net, we have a growing scientific story of the network of causal interactions within which mental activity occurs, interactions not only within the brain but among brain events and physical events external to the body. This wider causal ecology defines the causal substrate of consciousness. When philosophers like Fred Dretske (1981, 1995) collapse intentionality into the causal flow of physical "information" through

Mind World

environment and organism, they have limited their view of consciousness to its causal-ecological substrate.

A kindred view of mind is the widespread view that mind is a *function*, especially a computational function, of the brain. Causal or computational function is a higher-order property of brain activity, of the brain's mediation of inputs and outputs of the organism or system. These functional properties of brain belong to the physical, causal substrate of consciousness. Philosophers like Jerry Fodor (1975, 1994) and Daniel Dennett (1991) follow variations on the functionalist theme: Fodor identifies mental activity with a "language of thought" consisting in physical symbols processed in the brain, whereas Dennett identifies mind with computational brain function viewed from the "intentional stance" and consciousness with a particular function (producing "multiple drafts" of the brain's "story" about the world and itself). But functionalism is restricted to a specific view of the causal substrate of consciousness, overlooking the proper analysis of intentionality in the form and appearance of consciousness. (See D. W. Smith 1999a, calling for a wider ontology.)

The neural and causal grounding of consciousness is not the only sort of condition on which our conscious experience depends. As we have come to recognize, our cultural background also constrains or makes possible, in importantly different ways, the forms of intentionality we may enjoy. We cannot think about "naturalism" or "racism" or laissez faire economics except in a historical context in which other persons, other intentional subjects, have put forth and debated relevant issues. More basically, we cannot think as we do "in language" unless we have acquired a language, such as English. (See D. W. Smith 1999b on "background ideas.") Cultural preconditions, then, define a distinct region of the substrate of an act of consciousness. When philosophers like Richard Rorty (1979) reduce consciousness to philosophical "conversations" following Descartes, they restrict their view to the cultural substrate of our philosophical self-consciousness. But there is more to consciousness than what we say about it, even if our discourse shapes our awareness of our own experience. The form and appearance of our experience is distinct from its cultural background.

Still another kind of precondition of our human forms of intentionality is our biological heritage. Here is where the evolutionary biologist's point takes hold. I quoted Edward O. Wilson (1998) earlier, where I intimated that in the study of consciousness, physical and biological theory need to be developed in relation to phenomenology and ontology. Consider the role of biological evolution. The human organism – its nervous system, indeed its genome or overall genetic footprint – evolved in the natural environment of the planet Earth, in the planetary system of the star we call the sun, in this universe which has developed over some 12 billion years since the Big Bang. These natural conditions are preconditions for our forms of consciousness: for intentionality in the form of visual perception (by two eyes two inches apart on the front of a head), emotion (desire, fear, anger), cognition or thought (about water or fire or Plato), or volition (to run, using two primate legs rather than two differently advantaged lizardly aviary "legs"). Wilson is right that, in *some* sense, natural science will "explain" mind, consciousness, even the arts and humanities, even the natural sciences themselves as disciplines that have evolved in human cultures in the blink of cosmic time here on Earth. And philosophers such as Ruth Millikan (1984) and Daniel Dennett (1991, chap. 7) are right that, in *some* sense, intentionality and human consciousness will be "explained" by principles of biological evolution. These kinds of studies of consciousness, however, are limited to the biological *substrate* of consciousness. The intentional-subjective form and appearance of consciousness must be "explained" in different ways, in a phenomenological ontology that observes the three facets of consciousness.

The form and appearance of consciousness, featuring structures of intentionality, are simply different from its substrate, physical and cultural. This was the force of Husserl's long argument in the *Crisis* (1935–38); I resituate the claim, however, in terms of the three-facet distinction in categorial ontology. In recent philosophy of mind, John Searle (1992) has sharply separated the irreducibly subjective properties of consciousness and intentionality from their "background" of neural capacities, arguing thus against the prevailing physical-computational models of mind in cognitive science; again, I would resituate these differences within the three-facet ontology.

Here I want to stress that the *form* and *appearance* of consciousness are to be studied in their own right in phenomenology *cum* ontology, whereas the *substrate* of consciousness is to be studied in relevant disciplines in the physical, neural, biological, and cultural sciences. Given the three-facet distinction, we see what is wrong with the familiar ontological proposals in recent philosophy of mind, from reductive to eliminative materialism, from functionalism to computationalism to causal externalism to evolutionary psychosociobiology. These "naturalistic" theories are all looking for intentionality, qualia, subjectivity in the wrong places, in parts of the substrate of consciousness, rather than its form and appearance.

The Ontology of Substrate

To understand more clearly what counts as the substrate of consciousness, we need to address the ontological structure of substrate in general.

Every entity, we assume, has a *substrate*, initially defined as its "foundation" or "origin." But different entities may be founded or originated in very different ways, as our examples showed, and this fact may distract us from the basic ontological form involved. Fundamentally, the substrate of an entity consists of what it *depends* on for its existence, where A depends (ontologically) on B if and only if A could not exist unless B existed. This notion of dependence we must now consider.

For Aristotle, a quality in a substance cannot exist apart from the substance. Expanding on this notion, Husserl defined a dependent part as a part that cannot exist outside the whole, so that the part "requires foundation" by the whole.[2] But we need to separate part and dependence, because something may be dependent on an entity of which it is not a part. Thus, from Husserl's complicated scheme, we may distill the following definition of dependence:

> A *depends* or is *founded* on B if and only if A could not by essence exist unless B exists, that is, necessarily, by virtue of essence, A exists only if B exists.

We can also say that A is *grounded* on B, or B is the/a *ground* of A. While Husserl often used the term "foundation" (*Fundierung*) instead of "dependence" (*Unselbständigkeit*), I prefer to speak of dependence because the term "founded" suggests one-way dependence, yet two things may each depend on the other (neither could exist without the other).

There are different kinds of dependence: physical, biological, cultural, among others. But, on this analysis, the form of dependence is always the same: A could not exist unless B existed. Husserl adds the qualification "by essence," and Kit Fine (1995) further explicates dependence in terms of his own conception of essence. However, there are different kinds of conditions on which something can depend – for example, causal physical circumstances, evolutionary biological circumstances, social cultural circumstances. Moreover, there seem to be different kinds of "coulds," different modes of possibility or necessity – for example, what is physically necessary, biologically necessary, psychologically necessary, legally necessary. On the Husserl-Fine analysis, these differences may be attributed to essence: "necessarily, if A has essence E_A and B has essence E_B, then A exists only if B exists." These details of essence and necessity lie beyond the scope of the present study; for present purposes let us assume the basic account of dependence as defined earlier. (Remember that "essence" in Husserlian idiom means what something is, not what it is necessarily or "essentially.")

Many of the central problems of metaphysics involve ontological dependence. The *in re* theory of universals holds, with Aristotle, that a quality can exist only if instantiated in an individual (or "primary substance"). Causation may be analyzed in terms of dependence, as A is caused by B just in case A is physically dependent on B, or it is causally necessary that if B occurs then A occurs. Classical idealism holds with Berkeley, and realism denies, that material objects depend on their being perceived or otherwise projected by minds: that is to say, a material object, by its essence, can exist only if perceived or otherwise intended in consciousness. Kant's transcendental idealism holds that certain conceptual categories and certain forms of sensibility are the "necessary conditions of the possibility" of our knowledge of the empirical world around us. I would reconstrue this epistemological claim in the idiom of ontological foundation: our familiar forms of intentionality, or knowledge of objects in the world around us, could not occur unless we had acquired or inherited a certain repertoire of conceptual and sensory structures – that is, our familiar forms of intentionality depend or are founded on these conceptual-sensory structures. More recent philosophers – from Husserl (1913) and Heidegger (1927, as interpreted in Dreyfus 1991) and Wittgenstein (1953) to John Searle (1983, 1992) – have stressed the background social practices that condition our experience. Searle posits a "background" of acquired skills or practical capacities (realized in neural structure) on which our intentional states rest; as Searle puts it, these capacities "enable" our intentional states to represent what they do. I would explicate this enablement in the idiom of ontological foundation: our intentional states could not represent or "intend" what they do, our intentional relations could not obtain, unless these capacities existed in our "background." (Compare D. W. Smith 1989, chap. 6, and 1999b.)

An influential view in recent philosophy of mind and cognitive science holds that mental states are "supervenient" on physical states of the brain (Davidson 1970 and Kim 1994, 1998). Supervenience has been variously defined; however, the issues at stake here are covariance and dependence: the mental varies with the physical so that every mental state occurs along with an appropriate brain state and could not occur unless such a brain state occurred.[3] The core definition is, I think, best formulated as a doctrine of ontological dependence: every mental state is said to depend on some brain state in such a way that mental state types covary with physical state types.

David Lewis (1986) uses supervenience in the metaphysics of possible worlds. David Armstrong (1997) holds that a state of affairs (that an individual has a property or that two individuals stand in a relation) supervenes on its constituents (the individuals and universals involved). But Armstrong holds that supervenience brings into the world no new entities beyond those on which the supervenient ride: supervenience offers an "ontological free lunch," as Armstrong has put it.[4] Thus, a state of affairs would not be an entity in addition to the individuals and universals of which it is composed. And if mental activity supervenes on brain activity, there would be no new, distinctly mental entities beyond the physical entities, the brain states, on which mind supervenes. In the theory of ontological dependence outlined earlier, however, there is no free lunch. It does not follow from the preceding definition of dependence that A is not another entity beyond B. Quite to the contrary, in the cases that concern us, A and B are distinct entities between which a relation of dependence obtains, so that A can exist only if B exists.

An important but neglected form of dependence is one I discern in Alfred North Whitehead's ontology. In Whitehead's *Process and Reality* (1929), as commonly interpreted, the Aristotelian notion of substance, or enduring material object, is replaced by the notion of process (see Rescher 1996). However, Whitehead himself distinguished what I would call "ontological" becoming from temporal becoming. Every actual entity, Whitehead maintains, is constituted by its relations (of "prehension") to other actual entities in its process of "becoming," or "concrescence." But Whitehead distinguishes two types of "flux": the temporal flux by which a new entity emerges from others in the "transition" to the novel entity, and the "concrescence" of one entity formed out of many on which it depends. This latter notion I call *ontological* becoming (D. W. Smith 2000a). Whitehead's conception of concrescence I want to cast as a very fundamental form of dependence: one actual entity depends on a number of other entities in its existence, its "internal constitution." Indeed, the doctrine of logical atomism in Russell and Wittgenstein affords another instance of ontological becoming: a state of affairs is "logically" formed out of individuals and the properties or relations they bear. This complex entity depends ontologically on its constituents. But these are not like automobile parts, which are put together in a spatiotemporal relation to make the whole. A state of affairs is rather put together by "logical" relations of property instantiation. Whitehead did not adopt

Russell's "logical" atomism; instead, he elaborated an "atomism" where entities are put together by another kind of bond called "prehension."

Thus, *pace* Whitehead, the ultimate substrate of any "actual" entity or occasion, including an act of consciousness, is its "deep" ontological structure. In this way phenomenological structure rests on ontological structure (this, on the model adumbrated in D. W. Smith 2000a).

The Substrate of Consciousness

The substrate of consciousness is that aspect of an act of consciousness which consists not in how it is intentionally directed (its form), or in how it feels to its subject in inner awareness (its appearance), but in how it comes to be, how it originates ontologically, how it depends or is founded ontologically on various conditions in the surrounding world. In the normal course of events, I could not see or think or will what I do, in a given act of consciousness, unless certain background conditions obtained. That is, the act could not be directed as it is – from subject through content to object, with an inner awareness of the act built in – unless these background conditions obtained. The substrate of the act consists in its dependence on these background conditions. This pattern of relations among the substrate, form, and appearance of an act of consciousness may be depicted as in Figure 1.3 (where the T-bar stands for ontological dependence and the three facets are given a slightly different look).

As noted earlier, an act of consciousness may depend in different ways on a wide variety of things in the world. But we seek order in this variety. It seems that there are certain *kinds* of background conditions on which consciousness depends in systematically different ways. Thus, I propose, the substrate of an act divides into its patterns of dependence upon:

1. The psychological or personal history of the subject, which conditions his or her experience, making it more or less likely that she will see or think or feel or do things as she does.

FORM

background conditions |— < subject — act — content ———> object >

|

awareness-of-act

SUBSTRATE   APPEARANCE

FIGURE 1.3. Intentionality and the three facets of consciousness.

2. The cultural context or social history of the subject, which conditions his or her consciousness, affording her relevant concepts, values, and "forms of life" including language.

3. The neural activity in the subject's brain, including processes of computation in the neural network, which make possible this form of consciousness, and which in turn depends on appropriate physical and chemical processes.

4. The biological processes of evolution that allow the emergence of this form of consciousness, which thus condition the consciousness in still another way, and which in turn depend on the cosmological processes by which life itself evolved.

I believe these are the most basic, and fundamentally different, realms of being on which consciousness depends, in different ways. One might argue for a different organization of the substrate of consciousness, but my aim is first to organize the nature of an act of consciousness into form, appearance, and substrate, and then to look into the organization of the substrate. In fact, the preceding division of substrate – into psychological, cultural, neural, and evolutionary conditions of dependence – reflects a fundamental division of labor among the disciplines that have studied mind over the past century or two. Each of these four domains of study has developed as a rich and even revolutionary field: psychology, from psychoanalysis to cognitive science; cultural analysis, from social science to humanistic critical theory; neuroscience, becoming its own science in recent decades; evolutionary biology, again gaining salience in recent decades. If you will, this fourfold division in the substrate of consciousness is simply a classification of the chief empirical results, or "material ontologies," in the theory of mind. But these four areas in theory of mind are framed, here, by a distinction in formal ontology: the three-facet structure of form, appearance, and substrate, applying this structure to consciousness.

The philosophy of mind in recent decades has divided into theories that stress particular features of mind that I would systematize in the preceding division of the substrate of consciousness. Various naturalistic theories, we noted, stress causal, computational, neural, or biological conditions of those mental states we experience as consciousness. Culturalistic theories stress the origins of mental life in social conditions and human history. And psychologistic theories stress the contingent psychological origins of experience, rather than (as Husserl demanded) their meaning, or logically formed intentional content.

Phenomenological theories have, by contrast, distanced themselves here and there from precisely these tendencies. Now, all these dialectical cross-currents fall into place if we adopt a *systematic* ontology of the nature of consciousness: first distinguishing its form, appearance, and substrate, and then distinguishing these four regions of dependence within its substrate.

This diverse structure of the nature of mind comes into relief only as we look systematically at form, appearance, and substrate. And the special roles of ecological conditions of consciousness come to prominence only when we bring out the role of ontological dependence in the substrate of consciousness. We must not, then, *identify* the nature of consciousness with one of these types of conditions in its substrate, with brain, or computational function, or causal role, or cultural role, or evolutionary role. To depend on something, even deeply and fundamentally, is not to be identical with it.

The Return of Phenomenology

Consciousness is indeed what makes the mind-body problem difficult, when we look to the results of natural science – or indeed cultural theory. It is also what makes life worth living and philosophy, since Descartes and Kant and Husserl, so exciting.

We will not fully "understand" consciousness until we see how it fits into the structure of the world defined by quantum physics, evolutionary biology, cultural history, and even cosmology. But our understanding of consciousness must begin with our own experience, as Descartes began to see. Our understanding will begin with the structure of consciousness analyzed in phenomenology, and will go on to integrate the results of phenomenology with those of natural science and cultural analysis in a unified world picture framed by basic ontology.

Phenomenology will elaborate (part of) the *form* of consciousness in the structure of intentionality and will detail the *appearance* of consciousness in different forms of experience, including our inner awareness of experience. These analyses of form and appearance will interweave with logic, mathematics, computer science, and formal ontology, in analyses of forms of various things including consciousness. This complex of analyses will ultimately tie into analyses of the *substrate* of consciousness, comprising conditions under which the extant forms of experience are realized, conditions mapped out by the empirical natural sciences and the social or cultural sciences. And the structure of the world in which consciousness and its empirical background conditions obtain will be framed by basic formal ontology.

What is hard about understanding consciousness is getting our mind around all these different kinds of structure while keeping straight their differences. We do this as we delimit phenomenology and its kindred ontology.[5]

Notes

1. An extensive collection of work on consciousness in the literature of recent cognitive science is Block, Flanagan, and Güzeldere 1997. In these essays phenomenological issues arise quite often, including issues about consciousness, perception, qualia, content, intentionality, temporal awareness, and higher-order thought. Yet there are virtually no references, in this lengthy collection, to the literature of phenomenology itself, beginning with Husserl, where these issues have been analyzed with great illumination for a century. (William James appears, appropriately, the classical psychologist with a phenomenological nose.) Evidently, there is a cultural barrier at work. I shall not here specifically critique the recent literature in philosophy of mind *cum* cognitive science (Fodor, Dennett, et al.). My aim in this chapter is rather to assess issues of consciousness within the theoretical framework of phenomenology in the context of a wider ontology. In my "Mind and Body" (1995), I placed Husserlian theory within the space of ideas framed by recent philosophy of mind and cognitive science, including issues of reduction, functionalism, and the like. Meanwhile, issues of cognitive science are addressed in relation to phenomenology in Dreyfus 1982; Varela, Thompson, and Roesch 1993; and Petitot et al. 1999. My own essay in the latter volume, "Intentionality Naturalized?" (1999a), critiques two influential lines of analysis, articulated by Fodor and Dretske, from a phenomenological-ontological point of view.

2. Husserl's notion of founding (*Fundierung*), or dependence, was used in many later works after being developed in the third of his *Logical Investigations* (1900–1), elaborating a conception originating in Aristotle and used by Brentano. The Husserlian analysis is assayed in B. Smith 1982. A wider discussion of ontological dependence, amplifying Husserl's conception, is found in Simons 1987: chap. 8, pp. 290ff. Kit Fine explores an essence-based conception of dependence in Fine 1995. Thomasson (1999) develops a succinct model of a basically Husserlian conception of dependence, refining distinctions formulated by Roman Ingarden and applying the model to fictional objects as dependent on authors' and readers' intentional acts. D. W. Smith, (1989: chapter 6) sketches a ramified notion of dependence or ground, specifying different kinds of dependence (physical, psychological, etc.) and distinguishing kinds of dependence involved in direct awareness, or acquaintance, notably dependence on intentional content and different kinds of dependence on the context of one's experience. Simons (1987) similarly distinguished different kinds of ontological dependence, including logical presupposition.

3. See, as noted earlier, Kim 1994, especially the essay "Supervenience as a Philosophical Concept," and 1998.
4. See Armstrong 1989: 100, speaking there of internal relations as supervenient on their relata while being nothing over and above the relata.
5. The roots of the present essay are various. (1) I have assumed, broadly, a theory of intentionality as mediated by a meaning-content, the theory developed from Husserl in Smith and McIntyre 1982. (2) In D. W. Smith 1989, I developed a conception of the "ground" of intentionality, including physical, psychological, and cultural conditions on which intentionality depends; these grounds of intentionality are part of the substrate of consciousness, according to the three-facet ontology of consciousness discussed here. (3) A shorter ancestor of this essay is my "Ontological Phenomenology" (2000c). (4) The three-facet distinction in ontology, featured here, is used systematically in the formal phenomenological ontology of the PACIS project at Ontek Corporation: what I here call "facets" have been called "complements" in PACIS terminology. The distinction appears also in Simons and Smith 1993.

References

Armstrong, David M. 1989. *Universals: An Opinionated Introduction.* Boulder: Westview Press.

1997. *A World of States of Affairs.* Cambridge: Cambridge University Press.

Block, Ned, Owen Flanagan, and Güven Güzeldere, eds. 1997. *The Nature of Consciousness.* Cambridge, Mass.: MIT Press.

Chalmers, David. 1996. *The Conscious Mind.* Oxford: Oxford University Press.

Churchland, Patricia Smith. 1986. *Neurophilosophy.* Cambridge, Mass.: MIT Press.

Churchland, Paul M. 1995. *The Engine of Reason, the Seat of the Soul.* Cambridge, Mass.: MIT Press.

Davidson, Donald. 1970. "Mental Events." Reprinted in Davidson, *Essays on Actions, Causes, and Events.* Oxford: Oxford University Press, 1980.

Dennett, Daniel C. 1991. *Consciousness Explained.* Boston: Little, Brown.

Dretske, Fred. 1981. *Knowledge and the Flow of Information.* Cambridge, Mass.: MIT Press.

1995. *Naturalizing the Mind.* Cambridge, Mass.: MIT Press.

Dreyfus, Hubert L. (with Harrison Hall), ed. 1982. *Husserl, Intentionality and Cognitive Science.* Cambridge, Mass.: MIT Press.

1991. *Being-in-the-World: A Commentary on Heidegger's Being and Time, Division I.* Cambridge, Mass.: MIT Press.

Fine, Kit. 1995. "Ontological Dependence." *Proceedings of the Aristotelian Society* 95: 269–90.

Fodor, Jerry A. 1975. *The Language of Thought.* New York: Cromwell.

1994. *The Elm and the Expert: Mentalese and Its Semantics.* Cambridge, Mass.: MIT Press.

Føllesdal, Dagfinn. 1982. "Husserl's Notion of Noema." In Dreyfus 1982. First published in *Journal of Philosophy,* 1969.

Heidegger, Martin. 1988. *The Basic Problems of Phenomenology*. Translated by
    Albert Hofstadter. Revised ed. Bloomington: Indiana University Press, 1988.
    German original, 1975, from a lecture course in 1927.
Husserl, Edmund. 1969. *Ideas [Pertaining to a Pure Phenomenology and a Phe-
    nomenological Philosophy, First Book]: General Introduction to Pure Phenomenol-
    ogy*. Translated by W. R. Boyce Gibson. London: George Allen and Unwin;
    New York: Humanities Press. First English edition, 1931; German original,
    1913.
    1970a. *Logical Investigations*. Vols. 1 and 2. Translated by J. N. Findlay from
    the revised, second German edition. London: Routledge and Kegan Paul.
    New edition, edited with an introduction by Dermot Moran, and with a
    preface by Michael Dummett. London: Routledge, 2001. German orig-
    inal, 1900–1, revised 1913 (Prolegomena and Investigations I–V), 1920
    (Investigation VI).
    1970b. *The Crisis of European Sciences and Transcendental Phenomenology*. Trans-
    lated by David Carr. Evanston, Ill.: Northwestern University Press. Original
    manuscript from 1935–38.
    1991. *Ideas Pertaining to a Pure Phenomenology and a Phenomenological Philosophy,
    Second Book: Phenomenological Investigations of Constitution* [or as translated:
    Studies in the Phenomenology of Constitution]. Translated by Richard Rojcewicz
    and André Schuwer. Dordrecht: Kluwer Academic Publishers, 1991. German
    original, 1952, drafted initially in 1912, revised in 1915 and again in 1928.
ISO/IEC TR9007. 1987. Information Processing Systems – Concepts and Termi-
    nology for the Conceptual Schema and the Information Base. International
    Organization for Standardization (ISO), Geneva.
Kim, Jaegwon. 1994. *Supervenience and Mind*. Cambridge: Cambridge University
    Press.
    1998. *Mind in a Physical World: An Essay on the Mind-Body Problem and Mental
    Causation*. Cambridge, Mass.: MIT Press.
Lewis, David K. 1986. *On the Plurality of Worlds*. New York: Basil Blackwell.
Mayr, Ernst, and Peter D. Ashlock. 1991. *Principles of Systematic Zoology*. 2d. New
    York: McGraw-Hill.
Millikan, Ruth G. 1984. *Language, Thought, and Other Biological Categories: New
    Foundations for Realism*. Cambridge, Mass.: MIT Press.
Mohanty, J. N. 1985. *The Possibility of Transcendental Philosophy*. Dordrecht:
    Martinus Nijhoff Publishers.
Nagel, Thomas. 1974. "What Is It Like to Be a Bat?" *Philosophical Review* 4: 435–50.
Penrose, Roger. 1994. *Shadows of the Mind: A Search for the Missing Science of Con-
    sciousness*. Oxford: Oxford University Press.
Petitot, Jean, Francisco J. Varela, Bernard Pachoud, and Jean-Michel Roy, eds.
    1999. *Naturalizing Phenomenology: Issues in Contemporary Phenomenology and
    Cognitive Science*. Stanford, Calif.: Stanford University Press in collaboration
    with Cambridge University Press.
Rescher, Nicholas. 1996. *Process Metaphysics*. Albany: State University of New York
    Press.
Rorty, Richard. 1979. *Philosophy and the Mirror of Nature*. Princeton: Princeton
    University Press.

Searle, John R. 1983. *Intentionality.* Cambridge: Cambridge University Press.
——— 1992. *The Rediscovery of the Mind.* Cambridge, Mass.: MIT Press.
Sellars, Wilfrid. 1963. "Philosophy and the Scientific Image of Man." In W. F. Sellars, *Science, Perception and Reality.* London: Routledge and Kegan Paul; New York: Humanities Press.
Simons, Peter. 1987. *Parts.* Oxford: Oxford University Press.
Simons, Peter, and David Woodruff Smith. 1993. "The Philosophical Foundations of PACIS." Paper presented at the Sixteenth International Ludwig Wittgenstein Symposium, Kirchberg am Wechsel, Austria. August.
Smith, Barry, ed. 1982. *Parts and Moments.* Munich: Philosophia Verlag.
Smith, David Woodruff. 1989. *The Circle of Acquaintance.* Dordrecht: Kluwer Academic Publishers.
——— 1995. "Mind and Body." In Barry Smith and David Woodruff Smith, eds., *The Cambridge Companion to Husserl,* pp. 323–93. Cambridge: Cambridge University Press.
——— 1999a. "Intentionality Naturalized?" In Petitot et al. 1999: 83–110.
——— 1999b. "Background Ideas." Translated into Italian as "Idee di sfondo," Paradigmi (Estratto da PARADIGMI, Rivista di critica filosofica) (Rome), Anno XVII, no. 49 (January–April 1999): 7–37.
——— 2000a. "Consciousness and Actuality in Whiteheadian Ontology." In Liliana Albertazzi, ed., *The Origins of the Cognitive Sciences, 1870–1930: Theories of Representation,* pp. 1–30. Dordrecht: Kluwer Academic Publishers.
——— 2000b. "How Is Transcendental Philosophy – of Mind and World – Possible?" In Bina Gupta, ed., *The Empirical and the Transcendental: A Fusion of Horizons (Festschrift in Honor of J. N. Mohanty),* pp. 169–79. New York: Rowman and Littlefield.
——— 2000c. "Ontological Phenomenology." In Mark Gedney, ed., *Proceedings of the Twentieth World Congress of Philosophy,* vol. 7: *Modern Philosophy.* Bowling Green, Ohio: Philosophy Documentation Center, Bowling Green University.
Smith, David Woodruff, and Ronald McIntyre. 1982. *Husserl and Intentionality.* Dordrecht: D. Reidel.
Sokolowski, Robert. 2000. "Transcendental Phenomenology." In Mark Gedney, ed., *Proceedings of the Twentieth World Congress of Philosophy,* vol. 7: *Modern Philosophy.* Bowling Green, Ohio: Philosophy Documentation Center, Bowling Green University.
Thomasson, Amie L. 1999. *Fiction and Metaphysics.* Cambridge: Cambridge University Press.
Varela, Francisco, Evan Thompson, and Eleanor Roesche. 1993. *The Embodied Mind.* Cambridge, Mass.: MIT Press.
Whitehead, Alfred North. 1978. *Process and Reality.* Edited by David Ray Griffin and Donald W. Sherburne. New York: Free Press, Macmillan. First edition, 1929.
Wilson, Edward O. 1998. *Consilience: The Unity of Knowledge.* New York: Alfred A. Knopf.

2

The Cogito circa A.D. 2000

Abstract. What are we to make of the cogito (*cogito ergo sum*) today, as the walls of Cartesian philosophy crumble around us? The enduring foundation of the cogito is consciousness. It is in virtue of a particular phenomenological structure that an experience is conscious rather than unconscious. Drawing on an analysis of that structure, the cogito is given a new explication that synthesizes phenomenological, epistemological, logical, and ontological elements. What, then, is the structure of conscious thinking on which the cogito draws? What kind of certainty does the experience of thinking give one about one's thinking and about one's existence? What form of inference is the cogito, and what is the source of its validity and soundness? Does the cogito itself lead to an ontology of mind and body like Descartes's dualism? The discussion begins with Descartes's own careful formulations of some of these issues. Then the cogito is parsed into several different principles, the phenomenological principle emerging as basic. In due course the analysis sifts through Husserl's epistemology, Hintikka's logic (or pragmatics) of the cogito, and Kaplan's logic of demonstratives, as these bear specifically on the cogito.

Segue. In "Three Facets of Consciousness" we proposed that the nature of any entity, including an act of consciousness, divides into form, appearance, and substrate. Here we explore the *form* of consciousness – that of self-aware intentionality – through its *appearance* within an act, that is, through "inner awareness." We approach the analysis of inner awareness by a critical review of the traditional "cogito," drawing on Descartes's analysis but from today's perspective (without dualism and without incorrigibility). The results provide one type of justification for first-person methodology in the theory of mind. For phenomenology begins with our awareness of our own intentional experiences as we experience them from our first-person

For various comments on this essay and ideas expressed in it, I thank Ermanno Bencivenga, Alan Nelson, Calvin Normore, Martin Schwab, and Amie Thomasson.

perspective. (The *substrate* of consciousness, in brain and culture, is addressed in later essays. The terms "form," "appearance," and "substrate" are used here in the technical sense introduced in the preceding essay.)

There is something dead right, and very much alive, about the cogito: Descartes's epochal proclamation, *cogito ergo sum.* A conscious experience ("I am thinking") includes an awareness of itself, an awareness that makes it conscious and gives its subject immediate evidence of the experience and therein of his or her existence ("therefore I exist"). This principle is independent of Descartes's further doctrines (e.g., rationalism and dualism), and it survives the many refutations and deconstructions of the wider Cartesian corpus and their animus for the Cartesian *animus* (spirit). It tells us something crucial about consciousness. Yes, consciousness and some form of the self survive the transition from modern to postmodern philosophy in our century. Indeed, consciousness is something of the greatest importance not only to epistemology and psychology in the age of science but to ethics and social theory and practice in this perilous age on our planet. Without it, what are we?

Several different principles converge in the inference *cogito ergo sum,* including phenomenological, epistemological, and logical – semantic or pragmatic – principles that would ground the inference. Here I shall try to articulate some of these enduring principles, beginning with Descartes's own astute formulations, which I suspect are still not fully appreciated. I shall use "cogito" as an umbrella term for these several principles.

The Cogito in and out of (Descartes's) Context

I propose to study the cogito within the context of late twentieth-century philosophy, drawing on both continental and analytic traditions. I would start with a careful interpretation of Descartes's own formulations of the cogito, but I shall extract it from the wider context of Descartes's own times, concerns, and motivations.

Most philosophers – and mathematicians and scientists – would agree that Descartes's larger philosophical programs have not panned out. His beautiful course of argument in the *Meditations,* including one form of the cogito, has brought centuries of critique but has not yielded the foundation he sought for all knowledge, a foundation in the absolute certainty of reason, a certainty as strong or stronger than that he saw in mathematics. Nor has an alternative, purely rationalist epistemology emerged with the tremendous growth in mathematics, logic, and the natural sciences since his time, despite Descartes's seminal influence on these fields. His metaphysics of mind-body dualism seems passé, although the mind-body problem – so sharply defined by Descartes – is still very much with us, with no convincing solution, resolution, or dissolution in sight. And his theological assumptions do not seem of our time. Furthermore, his focus on the individual subject leaves unaddressed the social, cultural, historical side of the self – the subject of insightful work long after Descartes (for instance, in the writings of Heidegger, Sartre, Merleau-Ponty, and Foucault).

Nonetheless, Descartes's writings remain well worth close study. Not only do they show the genius and precision of a great philosopher at work at a certain point in history, but they remain admirably clear foils against which to view our subsequent heritage in epistemology, metaphysics, phenomenology, philosophy of mind and language, cognitive psychology, and artificial intelligence. In particular, his own discussions of the cogito are as precise as they are prescient of more recent developments in the logic and phenomenology of the cogito. Accordingly, we turn to Descartes's own presentation of the cogito.

Descartes's Cogito

Different principles are discernible in Descartes's discussions of the cogito. In the *Discourse on the Method* (1637)[1] he wrote:

[T]his truth "*I am thinking, therefore I exist*" was so firm and sure that all the most extravagant suppositions of the sceptics were incapable of shaking it [i.e., it was indubitable]. (CSM I, p. 127/AT, p. 32)

By contrast, in the *Meditations* (1641) he wrote:

[T]his proposition, *I am, I exist*, is necessarily [i.e., indubitably] true whenever it is put forward by me or conceived in my mind. (CSM II, p. 17/AT, p. 25; "proposition" translates the Latin *pronuntiatum*)

elaborating:

I am, I exist – that is certain [i.e., indubitable]. But for how long? For as long as I am thinking. (CSM II, p. 18/AT, p. 27)

Thus, the first formulation concerns an inference about thinking, whereas the second concerns not an inference but a principle about thinking. What did Descartes count as *thinking*? All the main types of mental activity (especially those we today call intentional). As he wrote in the *Meditations*: "What is [a thing that thinks]? A thing that doubts, understands, affirms, denies, is willing, is unwilling, and also imagines and has sensory perceptions" (CSM II, p. 19/AT, p. 28).[2]

Put in the present tense, Descartes's claim in the *Discourse* is that the inference, or inferential proposition, "I am thinking, therefore I exist" is indubitable (and hence indubitably true). Let us call this claim the "cogito inference principle." By contrast, Descartes's claim in the *Meditations* – in the first-person, present-tense – is that the proposition "I exist" is indubitable (hence, indubitably true) whenever I think it. Let us call this claim the "cogito principle." Of course, Descartes's program was to ground all our knowledge in propositions that are absolutely certain, where a proposition is certain if it cannot be doubted (it is indubitable, hence indubitably true). And the cogito was his first principle – formulated first as an inference in the *Discourse* and then as a principle in the *Meditations*.

To be precise, the two passages quoted from the *Meditations* articulate two different cogito principles. The first quotation formulates what we may call the *special* cogito principle.

> Whenever I am thinking the proposition "I exist," I am certain [I cannot doubt] that I exist, that is, the proposition "I exist" is indubitable (for me in so thinking).

The second quotation formulates what we may call the *general* cogito principle.

> Whenever I am thinking, I am certain that I exist, that is, the proposition "I exist" is indubitable (for me in so thinking)

– no matter what I am thinking. Behind both of these principles lies a different principle that Descartes seemingly took for granted in the *Meditations*, what we might call the protocogito principle.

> Whenever I am thinking, I am certain that I am thinking, that is, the proposition "I am thinking" is indubitable (for me in so thinking).[3]

(As in Descartes's own formulations, these principles involve second-order propositions – propositions about propositions. Thus, the special cogito principle concerns not only the epistemic proposition "I am certain that I exist" but the second-order, semantic proposition "the proposition 'I exist' is indubitable [for me . . .]." Differences will emerge between the claims here separated by "that is.")

Because the cogito inference principle does not specify any particular content of thinking, it aligns with the general cogito principle. But we might distinguish two cogito inferences too. The general cogito inference principle would say:

The inference "I am thinking, therefore I exist" is indubitable, that is, whenever I perform it, it is indubitable (for me in so performing it).

Whereas the special cogito inference principle would say:

The inference "I am thinking the proposition 'I exist,' therefore I exist" is indubitable, that is, whenever I perform it, it is indubitable (for me in so performing it).

It would be Descartes's claim, in either case, that what is indubitable is not only the *validity* of the inference but also the *truth* of its premise – the proposition "I am thinking" or "I am thinking the proposition 'I exist'" – and hence the *soundness* of the inference.

There can be no doubt as to which form of the cogito Descartes took to be fundamental for his purposes. Even in the *Discourse,* just a few lines before declaring the inference indubitable, he had put aside logic (reasoning, demonstrative proof) as open to doubt. Then in the *Meditations* what he claimed is indubitable is not the inference "I am thinking, therefore I exist," but the simpler proposition "I exist." At that point in his course of meditation, the evil-demon hypothesis had rendered logic dubitable. Moreover, in his *Replies* to critics of the *Meditations* (also 1641), he explicitly rejected logic as the ground of certainty even for the inference – at least the Aristotelian logic of his day, namely the theory of syllogism. He wrote:

When someone says "I am thinking, therefore I am, or I exist," he does not deduce existence from thought by means of a syllogism, but recognizes it as something self-evident by a simple intuition of the mind. This is clear from the fact that if he were deducing it by means of a syllogism, he would have to have had previous knowledge of the major premiss "Everything which thinks, is, or exists"; yet in fact he learns it from experiencing in his own case that it is impossible that he should think without existing. (*Second Replies*, CSM II, p. 100/AT, p. 140)

So it is "a simple intuition of the mind" that gives one certainty of one's existence: one learns the certainty of one's existence "from experiencing [it] in his own case." This certainty does not come from a logical intuition of the inference "Everything that thinks, exists, and I am thinking, therefore I exist" (an intuition recognizing the validity of this inference, or the logical truth of the inferential proposition). Nor, presumably, does it come from a logical intuition of the simpler inference "I am thinking, therefore I exist." The cogito inference may well be valid and its validity recognizable in a logical intitution, an intuition bearing the certainty of logic. But that is not the sort of certainty Descartes was seeking for one's own existence. Furthermore, Descartes clearly thought the cogito inference is not only valid but sound, and its lone premise "I am thinking" indubitable for the thinker. And the indubitability of the proposition "I am thinking" when one is thinking comes not from logic, or logical intuition, but from the experience of thinking, from a "simple intuition" of one's thinking.

Descartes's fundamental claim is explicit in yet another of the *Replies*:

It is true that no one can be certain that he is thinking or that he exists unless he knows what thought is and what existence is. But this does not require reflective knowledge, or the kind of knowledge that is acquired by means of demonstrations; still less does it require knowledge of reflective knowledge, i.e. knowing that we know, and knowing that we know that we know, and so on *ad infinitum*. This kind of knowledge cannot possibly be obtained about anything. It is quite sufficient that we should know it by that internal awareness which always precedes reflective knowledge. This inner awareness of one's thought and existence is so innate in all men that, although we may pretend that we do not have it if we are overwhelmed by preconceived opinions and pay more attention to words than to their meanings, we cannot in fact fail to have it. Thus, when anyone notices that he is thinking and that it follows from this that he exists, even though he may never before have asked what thought is or what existence is, he still cannot fail to have sufficient knowledge of them both to satisfy himself in this regard. (*Sixth Replies*, CSM II, p. 285/AT, p. 422)

So, according to Descartes, even when one performs the (general) cogito inference, one's certainty of one's existence comes from one's "inner awareness of one's thought and existence." It does not come from "reflective knowledge" or, more specifically, from logical inference ("the kind of knowledge that is acquired by means of demonstrations"). In short, the source of certainty for one's knowledge of one's own existence – and even for the validity of the cogito inference – is not *logical* but *phenomenological* (to use a term not yet available in Descartes's day). And that is the point of the general cogito principle formulated here.[4]

Having explained the cogito principle in this way, Descartes might have abandoned the cogito inference as at least misleading. Yet he reiterated the cogito inference in the *Principles of Philosophy* (1644): "[I]t is a contradiction to suppose that what thinks does not, at the very time when it is thinking, exist. Accordingly, this piece of knowledge [this inference: French version] – *I am thinking, therefore I exist* – is the first and most certain of all to occur to anyone who philosophizes in an orderly way" (CSM I, p. 195/AT, p. 7). Although this formulation is less precise than the earlier ones (for one thing, it is cast in the third person), Descartes explicitly endorses the certainty of the cogito inference per se. This should tell us that he does not want to eliminate the inference in favor of the cogito principle. And the passages quoted previously tell us why: the principle underlies the inference, in that the principle explicates the source of certainty Descartes claims for the inference itself.

This prereflective inner awareness of one's thinking – the source of certainty in the cogito – is a "perception" of one's thinking, in a generalized sense of perception. As Descartes wrote in *Passions of the Soul* (1649, a year before his death):

> Our perceptions are . . . of two sorts: some have the soul as their cause, others the body. Those having the soul as their cause are the perceptions of our volitions and of all the imaginings or other thoughts which depend on them. For it is certain that we cannot will anything without thereby perceiving that we are willing it. And although willing something is an action with respect to our soul, the perception of such willing may be said to be a passion in the soul. But because this perception is really one and the same thing as the volition, . . . we do not normally call it a "passion," but solely an "action." (CSM I, pp. 335–36/AT, p. 343)

Thus, one's inner awareness of one's thinking is "really one and the same thing as" the thinking itself. Although Descartes's action-passion distinction is not easy to explicate, his words are echoed in Sartre's doctrine of the prereflective cogito, the immediate knowledge one has of one's own consciousness, which Sartre says is "constitutive" of consciousness.[5]

What should we make today of these several principles that course through Descartes's discourse about the cogito – from the special and general forms of the cogito principle to the cognate cogito inferences? I shall approach these principles from different directions, seeking plausible grounds for them in relevant domains of phenomenology and logic (including semantics and pragmatics) and addressing the more familiar issues in epistemology and ontology from those perspectives.

The Phenomenology of the Cogito

The heart of the cogito is the *phenomenological* principle that we have an immediate *awareness* of our own current experience. This type of awareness gives one immediate *evidence* – and, so far as one can tell, immediate knowledge – of the fact that one is so thinking and the fact that, in so thinking, one exists. Whether this type of evidence yields the kind of *certainty* Descartes sought – namely indubitability – is a further issue. This phenomenological principle I propose to explicate in terms of a structural analysis of this "inner awareness" of experience.

The general, not the special, cogito principle has fundamental phenomenological force. Let us recast the general cogito principle in the following form, incorporating the protocogito principle, thus making explicit one's certainty not only of one's existence, but of one's thinking and indeed the content of one's thought:

> When I am thinking "*p*," I am certain that I am thinking "*p*" and that I exist.

We must restrict the principle, however, to *conscious* thought. Thus, the general cogito principle should say:

> When I am consciously thinking "*p*," I am certain that I am thinking "*p*" and that, in so thinking, I exist.[6]

(Because Descartes included under "thinking" all intentional mental activities, the principle should be generalized by replacing "thinking" with a general term ranging over all intentional mental activities. However, for easy reading, in the spirit of Descartes, we will retain the term "thinking.") Descartes did not appear to recognize *unconscious* mental activities. Indeed, the theory of the unconscious was to develop only later, especially in Freud's psychoanalysis and in today's cognitive psychology. The cogito, however, applies only to conscious mental activities: I have an immediate *awareness* – and therewith immediate evidence and arguably a kind of certainty – of my own mental activity only if it is a conscious mental activity. Indeed, the inner awareness I have of my current mental activity is precisely what makes it a *conscious* mental activity. (Or so the present author has argued on another occasion.)[7] This conception of consciousness was no stranger to Descartes, even if the unconscious was. As quoted earlier, Descartes himself explained the cogito principle in terms of "inner awareness of one's thought and existence," which is an "intuition of the mind," a "perception" of one's own thinking. Moreover, in the *Principles of Philosophy* (1644) he wrote, "By the term 'thought,' I understand everything which we are aware of as happening within us, in so far as we have awareness of it" (CSM I, p. 195/AT, p. 7). In E. M. Curley's translation, this reads: "all those things which, we being conscious, occur in us, in so far as the consciousness of them is in us," drawing on the Latin *immediate conscii simus.*[8]

Thus, the cogito principle is fundamentally a principle about the nature and structure of consciousness per se, that is, the quality of being conscious (being a conscious, as opposed to unconscious, mental activity). Elsewhere I have proposed an analysis of the defining structure of consciousness.[9] On that proposal, a conscious mental activity has the following *phenomenological* structure (staying with the Cartesian verb "think"):

Consciously I now am thinking "*p.*"

The structure of the quality "consciously" may be unfolded further as follows:

Phenomenally in this very experience I now am thinking "*p.*"

Here we distinguish four relevant phenomenological features in a conscious mental activity: the *phenomenal* quality, or "quale," of the experience; the *reflexive* character "in this very experience"; the egocentric character "I [think]"; and the *temporal* character "now." The phenomenal quality of an experience is the subjective character of what it is like as experienced.[10] The reflexive character of an experience is its character of *indicating itself* as it is experienced. The egocentric character of an experience is its character of *indicating its subject* as having the experience. And the temporal character "now" indicates the temporal position of the experience within the stream of retained past and expected impending experiences.[11] My *awareness of my thinking* "*p*" lies in the reflexive character of my thinking, and my *awareness of myself* in thinking lies in the egocentric character of my thinking. Fundamentally, it is this reflexive awareness of my thinking that makes my thinking conscious, and this awareness is qualified further by the accompanying egocentric, phenomenal, and temporal characters.

Importantly, these four features of an experience of thinking are not part of "what" is thought, but part of "how" something is thought. What is thought – the content of the act of thinking – is a *proposition*; how it is thought is a *modality* of thinking. These four "modal" features of an experience are proper parts of the experience. Thus, they are present in the experience prior to any reflection on the experience, which takes place in a separate, further mental act (either at the same time or shortly thereafter). They are thus, in Sartre's phrase, "prereflective." Indeed, as quoted earlier, Descartes said inner awareness is not "reflective knowledge" of one's thought. Accordingly, if *introspection* is a separate act of phenomenological reflection, as I think it is usually conceived, then inner *awareness* is distinct from introspection.[12]

In terms of this analysis of phenomenological structure, we can now explicate the general cogito principle. The *principle of awareness* – as we may now introduce it – says:

> When I am consciously thinking "*p*," I am aware of my thinking "*p*" and therein of myself.

My awareness of my experience lies in the *reflexive* character of the experience, and my awareness of myself lies in the accompanying *egocentric* character of the experience. These two forms of "inner" awareness ground the general cogito principle, for they carry my immediate *evidence* of the fact that I am so thinking and therewith the fact that I exist in so thinking. Thus, the principle of awareness may be formulated alternatively as:

> When I am consciously thinking "*p*," I am aware and it is thereby evident to me that I am thinking "*p*" and that I exist (in thinking "*p*").

And if the evidence of inner awareness amounts to *certainty*, then we have here the general cogito principle that Descartes enunciated.

Let us be clear about the role of qualia in "evidence." We have focused on thinking in the sense of judgment, but Descartes included under thinking all intentional mental activities, including sense perception. Now, a visual perception has an *evidential* character that is grounded partly in the *phenomenal* character, or quale/qualia, of the perception. However, one's sensory evidence of the world before one is different from the evidence of one's experience that is part of inner awareness. And it is the latter kind of evidence that pertains to the cogito principle.

The *temporal* character of experience – "I *now* am thinking '*p*'" – complicates the issues of the cogito in an interesting way. For this character rings in one's sense of personal identity through time. As Husserl observed,[13] drawing on his analysis of time consciousness, the "I" of whose existence one is certain in performing the *ego cogito* is not a momentary I, but a temporally extended I. Likewise, the act of thinking of which one is certain is a temporally extended process. In our terms: in my inner awareness of my current experience, I am aware of my experience as unfolding through time and I am aware of myself as an enduring subject of that experience. The span of time so experienced is, however, relatively short: it is what William James called the "specious present"; only if Husserl was right does it have a somewhat greater spread owing to the structure of time consciousness in one's fading "retentions" of the passed phases of one's experience.

Let us look more closely now at the problem of evidence in the cogito.

The Epistemology of the Cogito

The uncertainty in the general cogito principle, I submit, concerns the *certainty* gained in inner awareness. *What kind of certainty* does one have of one's own thinking and existence *in virtue of* one's inner awareness of one's experience? Descartes sought absolute indubitability. And through the indubitability of the cogito he sought the defeat of skepticism (the skepticism he had formulated in the First Meditation, culminating in the demon hypothesis) and ultimately a rationalist foundation for all knowledge. Here I want to follow a road less traveled, seeking the "certainty" of the cogito in the phenomenology of evidence.

Husserl understood *evidence* (*Evidenz*, self-evidence) as a phenomenological quality in various types of experience, and he distinguished different *types of evidence*, including the evidence of phenomenological reflection (Husserl's mode of introspection, guided by his methodological strictures).[14] Similarly, we might recognize in inner awareness its own kind of evidence. Sensory perception is an evident presentation of one's physical surroundings. The quality of sensory evidence is epistemically primitive: our judgments about our surroundings are based on sensory perceptions in virtue of their characteristic evidence, and such judgments are revisable in light of further sensory evidence and even further beliefs. But there is nothing to say about sensory evidence itself beyond a phenomenological description of perception as carrying evidence in varying degrees, depending, say, on how well one sees something. (What we normally call sensory evidence, then, comprises the propositions that are contents of judgments based immediately on sensory perceptions.) Similarly, inner awareness is an evident awareness of one's current mental activity. This kind of evidence too is epistemically primitive: our judgments about our own mental activities are based on the evidence in inner awareness, yet about this kind of evidence itself we can only offer phenomenological description, including strength or degree. The analogy with perception is not new. As noted earlier, Descartes himself used a generic notion of perception, which explicitly included not only sensory perception but perception of one's thinking. The outstanding question is about revisability. Are introspective judgments – judgments about our experience based on the evidence of inner awareness – *revisable* in light of further inner awareness, or further evidence in sensory perception, or in light of auxiliary beliefs? Is the evidence in inner awareness so strong that the deliverances of inner awareness are always and forever unrevisable?

It is commonly said that Descartes assumed the "incorrigibility" of the mental, this interpretation of Descartes deriving from Gilbert Ryle's *The Concept of Mind* (1949).[15] As Ryle put it, Cartesian doctrine holds: "The states and operations of a mind are states and operations of which it is necessarily aware, in some sense of 'aware,' and this awareness is incapable of being delusive.... It is part of the definition of their being mental that their occurrence entails that they are self-intimating" (p. 158). Because not all mental states are conscious (as Ryle observed *contra* Descartes), this claim should be restricted to the conscious ones. Then the claim is that (1) every conscious mental state includes an awareness of itself and (2) that awareness is infallible, or "incapable of being delusive." This claim is the conjunction of two principles, the first being essentially the principle of awareness formulated previously (for the case of thinking):

When I am consciously thinking "p," I am aware of my thinking "p."

The second, which we may call *the principle of infallibility (of inner awareness)*, would say (for the case of thinking):

When I am aware of my thinking "*p*," I am in fact consciously thinking "*p*."

This principle is just the converse of the principle of awareness, and it is equally plausible: consciousness and awareness go hand in hand (in normal adult human consciousness). But this is an ontological claim. And "incorrigibility" (uncorrectability, unrevisability) ought to be an epistemological feature of introspective, phenomenological judgments about conscious thoughts and experiences. So our question is, What kind of evidence does inner awareness give to introspective judgments? Are they so secure in their evidence that they could not reasonably be revised?

Descartes's answer we know: it is his Second Meditation. When I am consciously thinking "*p*," the evidence in my awareness of my so thinking is so strong that I cannot reasonably doubt that I am so thinking – even under the hypothesis of an evil demon who places the thought in my mind. Thus, when so thinking I could not reasonably revise my judgment that I am so thinking (insofar as I could form this judgment at the same time, in the presence of that thinking).

What did Husserl say when he addressed the cogito in his *Cartesian Meditations*? Husserl said that the *evidence* in our reflection on our experiences has the three characters of certainty (we do not doubt these judgments), adequacy (there are no hidden aspects of pure consciousness), and even apodicticity (it is "unimaginable" that we could doubt these judgments).[16] For Husserl, then, the apodicticity of judgments made in phenomenological reflection – based, let us add, on the evidence in inner awareness – would seem to rule out any reasonable revision of these judgments. And yet, the phenomenological character of apodicticity does not entail the ontological character of infallibility: although it may seem unimaginable to me – in light of my awareness of thinking "p" – that there could be reason for me to doubt that I am now thinking "p," in fact I may not be so thinking.

Indeed, there is much to know about our experiences that is not accessible to phenomenological reflection but is the business of empirical psychology, from psychoanalysis to experimental cognitive psychology to neuroscience and psychobiology. The epistemology of inner awareness and introspection should thus allow that judgments of introspection are, like all things human, fallible and, in principle, open to revision in light of appropriate evidence from further introspection and from appropriate empirical investigation, say, about one's current brain state. By the same token, empirical theories of mind – in psychobiology and cognitive science – must accommodate the evidence of inner awareness and introspection.

A proper epistemology of inner awareness and introspection would assay the types and grades of evidence therein. Such a project is well beyond the scope of this chapter and probably beyond the state of current knowledge. But note how very strong is Descartes's criterion of certainty in the cogito. Not only is indubitability the measure of certainty (whereas Husserl allows a weaker form of certainty than apodicticity or indubitability). But this indubitability, for Descartes, is measured by performing a thought experiment in which one imagines a wildly possible circumstance: an evil demon orchestrating one's experience. This thought experiment defines something like a logical possibility, raising the question of what precisely is the *logic* of the cogito and what kind of *logical possibility* the demon scenario exemplifies in regard to the proposition "I am thinking."

The Logic of the Cogito

Any appraisal today of the logic of the cogito should begin with Jaakko Hintikka's perceptive study, "*Cogito, Ergo Sum*: Inference or

Performance?" (1962).[17] While Hintikka's exact proposal – the so-called performative version of the cogito – is neither the most plausible reading of Descartes's texts nor the most compelling form of the cogito, his subtly crafted essay separates many of the key issues concerning the logic of the cogito. Here we shall pursue those and related issues in light of more recent results in the phenomenology of inner awareness and the logic of indexicals such as "I."

Hintikka rightly observed that if *cogito ergo sum* expresses a logical inference, it cannot be an inference of the form "*a* is thus and so, therefore *a* exists," or an implication of the form $Fa \rightarrow \exists x(x = a)$ in quantification logic. For these principles merely formalize the presupposition that singular terms in the given language refer to existent entities. And the question is whether (with Descartes perhaps) we can simply assume a logic with this presupposition, as there are logics that are free of this presupposition (they have come to be called "free logics").[18] Moreover, even if this form of inference is valid, it does not help to explain what is special – or what Descartes thinks is special – in *cogito ergo sum*, which concerns thinking. Similarly, on an ontological level, critics have long charged that Descartes tacitly assumed a Scholastic ontology of substance and attribute, reflecting a similar form of inference: if something has an attribute, then it is a substance and so exists. But this interpretation does not explain what is special about the cogito and its place in Descartes's epistemology. And similarly on a semantic level, Jerrold Katz has argued at length for an updated notion of analyticity, given which *cogito ergo sum* expresses an analytic entailment, as the concept or sense "thinking" entails the sense "exist."[19] But, again, this interpretation does not deal with what is special about the cogito in Descartes's program.

Hintikka proposed, accordingly, that *cogito ergo sum* does not express an inference of the usual sort but rather reflects the "performative" character of the statement made by uttering the sentence "I exist," a statement whose denial is "curiously pointless" (p. 58) or self-defeating (p. 60, n. 23). More precisely, where *p* is a sentence and *a* a singular term, Hintikka defined *p* as *existentially inconsistent for the person referred to by a to utter* if and only if the sentence "*p*; and *a* exists" is inconsistent (in the ordinary sense) (p. 57). The negation of such a sentence is then *existentially self-verifying* for the speaker (p. 60). Thus Hintikka's appraisal:

Descartes realized, however dimly, the existential inconsistency of the sentence "I don't exist" and therefore the existential self-verifiability of "I exist." *Cogito, ergo sum* is only one possible way of expressing this insight. Another way actually employed by Descartes is to say that the sentence *ego sum* is intuitively self-evident.

We can now understand the relation of the two parts of the *cogito, ergo sum* and appreciate the reasons why it cannot be a logical inference in the ordinary sense of the word. What is at stake in Descartes's dictum is the status (the indubitability) of the sentence "I am." . . . Contrary appearances notwithstanding, Descartes does not demonstrate this indubitability by deducing *sum* from *cogito*. On the other hand the sentence "I am" ("I exist") is not by itself logically true, either. Descartes realizes that its indubitability results from an act of thinking, namely from an attempt to think the contrary. The function of the word *cogito* in Descartes's dictum is to refer to the thought-act through which the existential self-verifiability of "I exist" manifests itself. . . . It serves to express the performatory character of Descartes's insight; it refers to the "performance" (to the act of thinking) through which the sentence "I exist" may be said to verify itself. (pp. 61–62)

So, on Hintikka's analysis, there is no logical problem with the *sentence* "I do not exist"; it is not logically false, nor is the sentence "I exist" logically true. But there is a pragmatic problem with the *statement* made – the speech act performed – by uttering the sentence "I do not exist": "[N]obody can make his hearer believe that he does not exist by telling him so" (p. 58). Similarly, no one can make himself believe that he does not exist by *thinking* he does not, by thinking the proposition "I do not exist" (if we may here import the notion of propositions as thought contents). "This transition from 'public' speech-acts to 'private' thought-acts," Hintikka adds, "does not affect the essential features of their logic" (p. 59). Thus, to interpolate a bit, the *proposition*, or thought content, "I do not exist" is not logically false, nor is the thought-content "I exist" logically true; but the *act* of thinking "I do not exist" is self-defeating, and hence the act of thinking "I exist" is self-verifying. Moreover, Hintikka assumes, the "performance" through which the *sentence* "I exist" "may be said to verify itself" is the act of *thinking* "I exist": it is only through the act of thinking "I exist" that the speech act of *stating* "I exist" – stating that one exists by uttering the sentence "I exist" – is verified; and only in that way is the sentence "I exist" self-verifying. Finally, the *proposition* "I exist" is self-verifying insofar as its truth is (in Hintikka's phrase) *intuitively self-evident* in performing the act of thinking "I exist." And thus is the proposition "I exist," *pace* Descartes, indubitable for the subject in the act of so thinking.

If Hintikka's analysis is on the right track, the *logic* of the cogito concerns not the validity or soundness of the form of *inference* "I am thinking, therefore I exist," but the self-verification of the act of *thinking* "I exist" and on that basis *stating* that one exists by uttering the sentence "I exist." So the purview of *logic* must be extended to include not only sentences (in syntax and semantics), but also speech acts (in pragmatics) and even thought acts (in phenomenology). I believe this much is quite right and consonant with the history of logic.[20] As noted earlier, however, there are really two cogito principles (plus two cogito inference principles) to which this analysis must be applied, and Hintikka's exact claim applies to only one.

The special cogito principle, enunciated in Descartes's Second Meditation (quoted earlier), says:

> Whenever I am thinking the proposition "I exist," I am certain (I cannot doubt) that I exist.

This is precisely the principle Hintikka's analysis underwrites. By contrast, the general cogito principle, enunciated in Descartes's *Replies* (quoted earlier), says:

> Whenever I am thinking [thinking anything at all], I am certain (I cannot doubt) that I exist.

This is the principle explicated earlier, based on the proposed phenomenology of inner awareness of thinking. Can this general cogito principle be underwritten by the logic of the cogito, in terms of the broader conception of logic suggested by Hintikka's appraisal of the cogito? To this we shall turn shortly, joining the logic and the phenomenology of the cogito. But notice that these two principles have very different grounds. On the one hand, there is something specific to the proposition "I exist" that makes the act of thinking "I exist" *self-verifying*, and so grounds the special cogito principle. On the other hand, there is something general about conscious thought (about inner awareness of one's thinking) that makes one *certain* of one's thinking and of one's own existence in thinking, and so grounds the general cogito principle.

Regarding the special cogito principle, we should distinguish two "performative" characters that might ground the principle. Hintikka says the transition from public speech acts to private thought acts does not affect their logic, that is, their performative characters. But is that right? If I say to you, "I exist," my assertion is curious and self-verifying because I am already visually before you, so what is the point of my saying that – why will your hearing me say that help to get you to believe I exist when you already see me before you? However, if I think to myself "I exist," my thought is curious and self-verifying, not because I am in a social circumstance, conversing with another, trying to convey beliefs by speaking, but because I already have an inner awareness of my thinking and therein of my existence – so what's the point of thinking (judging) "I exist"? The pragmatics of these two acts, the overt speech act and the inner mental act, are very different. If Hintikka's analysis is correct for the public speech act, another analysis is still needed for the private thought act – namely, analysis of inner awareness itself, not an analysis of the pragmatics of conversation. And, indeed, as Hintikka himself suggested, the self-verification of the speech act is grounded in the self-evidence of the thought act.

Hintikka's analysis addresses the statement "I exist." But what of the *inference* "I am thinking, therefore I exist"? The cogito, after all, was supposed to *prove* something (as my colleague Ermanno Bencivenga remarked to me). Logic would demonstrate the validity of the inference, on any occasion of performance, by showing somehow that the premise could not be true and the conclusion false on that same occasion. Its validity, however, is not grounded simply in syntactically defined rules of inference (e.g., $Fa/ \therefore \exists x(Fx)$). The "self-validating" character of the inference – whence its validity – is not a matter of proof theory, but a matter of semantics or pragmatics grounded in phenomenology, as we shall see later. Further, the truth of the premise "I am thinking" – hence the soundness of the inference – would be a matter of logic only if this premise were somehow a *logical* truth, and its truth, as we will find shortly, is warranted by phenomenology rather than logic as usually conceived.

There is, moreover, an unusual epistemic feature of Descartes's use of the cogito inference (general or special). Descartes used the cogito inference in his quest for certainty. Thus, in the inference "I am thinking, therefore I exist," what Descartes sought to preserve – in the move from "I am thinking" to "I exist" – was not merely truth but *certainty*, or indubitability: the indubitability of my thinking (for me in thinking) is transferred to the indubitability of my existence (for me in so thinking). In this respect (as Calvin Normore wisely observed in conversation), Descartes's conception of inference in the cogito was different from our twentieth-century notion of a truth-preserving movement of thought. Or, rather, Descartes asked more of this particular inference than truth preservation, and the indubitability he sought to preserve derives from the character of evidence in inner awareness of thinking.

Many of Descartes's readers have thought that introspection – or better, as argued earlier, inner awareness – is the source of certainty in Descartes's cogito. Hintikka played down the role of introspection in the cogito (pp. 63ff.), stressing the logic. But that was before he had discovered phenomenology. In another context he has since quipped, "There is often more logic than phenomenon in phenomenology." Turning to the cogito and his appraisal of its logic, however, we might better say, "There is more phenomenon than logic in the cogito." The phenomenon of thinking – the act of thinking as it is experienced – is the ground of the logic of the cogito. Better still: there is logic in the phenomenology and phenomenology in the logic of the cogito! And there is linguistics, too, in the range of speech act theory and language game theory.

Thus, to appraise the *logical* force of the cogito (principle and inference, special and general), we have to appraise the several lines of force – phenomenological, semantic, and pragmatic – in the relevant *mental acts, speech acts, sentences*, and *propositions* (thought contents). These include:

> The sentences "I exist" and "I am thinking, therefore I exist."
> The speech acts of asserting, "I exist," and inferring, "I am thinking, therefore I exist."
> The propositions "I exist" and "I am thinking, therefore I exist."
> The mental acts of thinking "*p*," thinking "I exist," and thinking "I am thinking, therefore I exist."

The Semantics (or Formal Pragmatics) of the Cogito

A lot of water has flowed under the bridge of philosophical logic since Hintikka's essay on the cogito. Along came possible-worlds semantics, including Hintikka's formulations of the logic and semantics of sentences ascribing intentional attitudes; then came Richard Montague's extension of possible-worlds semantics to pragmatics, giving formal models of context-dependent aspects of meaning and reference; and then, along with the neo-Russellian theory of "direct" reference, came David Kaplan's logic – formal semantics or pragmatics – of demonstrative pronouns like "that" and indexical words like "I," terms whose reference depends on the context of utterance.[21] Noting the indexical elements in "I exist" and "I am thinking," we might now try to explicate the logic of the cogito in terms of a formal semantics (or pragmatics) of indexicals – enlightened by the phenomenology of thought and inner awareness. Specifically, in terms of such a semantics we can explicate the self-verifying character of a speech act of asserting, "I exist" – or indeed "I am thinking, therefore I exist." And in terms of a parallel "semantics" of thought, we can explicate the self-verifying character of a mental act of thinking "I exist" – or "I am thinking, therefore I exist."

Kaplan's semantics for indexicals ramifies Frege's scheme of sense and reference and reintroduces a neo-Russellian notion of "singular propositions."[22] In place of Fregean "sense," Kaplan introduces two levels of meaning called "character" and "content." Take, for instance, the sentence "I am tall." The *character* of the sentence – something like its role or rule of use – specifies for each *context* (or occasion) of utterance the *content* of the sentence in that context, and the *content* of the sentence in a given context is, for Kaplan, the "singular proposition" consisting of the individual whom "I" refers to in that context and the property of being tall – in effect, the putative state of affairs that said referent is tall. Of course, the character of the sentence "I am tall" depends on the character of the indexical "I," which specifies for each context of utterance the content of "I" in that context, which is simply the individual who utters "I" in that context. Indeed, in any context "I" *refers* to the individual who utters "I" in that context; for Kaplan, "I" refers "directly," in that the referent is fixed by context without the mediation of a Fregean "sense," and so the "content" of "I" in a given context is simply its referent in that context. (Originally, Kaplan modeled character as a function from contexts to contents, and content as a function from possible worlds to extensions. *Nota bene*: I have used the terms "content" and "proposition" differently, meaning intentional contents of thoughts or experiences, and I shall resume that usage later.)

What about the cogito? In Kaplan's semantics, the sentence "I exist" is true whenever it is asserted, that is, in any context of utterance. If you will, it is "analytically" true, true by virtue of the level of meaning called "character," and in that sense its utterance is "self-verifying." That is: the character of "I exist" specifies for any context of utterance the "singular proposition" consisting of the individual uttering the sentence in that context and the property of existence;[23] and, whatever the context, that "proposition" is true – since, by hypothesis, the context of utterance includes the speaker, who must exist in that context in order to perform the utterance of the sentence "I exist." Thus, Kaplan's semantics underwrites a *linguistic* version of Descartes's special cogito principle: "The sentence 'I exist' is necessarily [i.e., analytically] true whenever I pronounce it." However, the inference "I am thinking, therefore I exist" is not, in Kaplan's semantics, "necessarily true" in every context of utterance (semantically valid and sound) – unless it is assumed, say, that in every context the speaker is thinking and thinking entails existence, which begs the question of the cogito. Nor does the semantics address that the speaker is thinking and thinking entails existence, which begs the question of the cogito. Nor does the semantics address the speaker's *certainty* of his or her existence in uttering "I exist" or "I am thinking, therefore I exist"; indeed, Kaplan's semantics per se sheds no light on the epistemological or phenomenological features of the cogito (nor was it intended to).

Thus, we need a semantics formulated with an eye to phenomenology, to intentionality theory, to the "semantics" of thought. We need a semantics that posits levels of meaning coordinate with the intentional structures of thought, namely, the *content* and *object* of thought. Accordingly, let us sketch an *intentional* semantics (or pragmatics) distinguishing four levels of meaning called character, sense, intension (or intention), and extension.[24] Take, again, the sentence "I am tall." The (*semantic* or *pragmatic*) *character* of the sentence – its rule of use – specifies for any context of utterance its sense and its intension in that context. Its *sense* in a given context is the proposition expressed in that context, namely the thought content "I am tall." Its *intension* (or *intention*) in the given context is the state of affairs asserted (or "intended") in that context, namely the singular state of affairs consisting of the referent of "I" in that context and the property of being tall. Its *extension* in that context is its truth value in that context. And it is *true* in a given context of utterance if and only if the state of affairs asserted by uttering it in that context is actual: so its character determines its truth conditions. Thus, the proposition expressed by saying "I am tall" (the sense) would be the *content* of the speaker's thinking "I am tall" on that occasion, while the state of affairs asserted (the intension) would be the *object* of the speaker's so thinking.

Of course, the meaning of "I am tall" depends on the meaning of "I," that is, its character, sense, intension, and extension. The character of "I" specifies for each context of utterance the sense and intension of "I" in that context: in any context of utterance, "I" expresses the sense "I," which embodies the mode of presentation "I" (more on that shortly), and "I" refers to (or "intends") the person uttering "I" in that context, who is thus the intension of "I" in that context. And the extension of "I" in that context is (also) the referent in that context. *Contra* Kaplan, in this intentional semantics, "I" refers "directly" not because it has no "sense" and is fixed by context, but because its reference is geared to the speaker's "direct" awareness of him- or herself in speaking – that is, because in any context of utterance, "I" expresses the sense "I," which embodies the "direct" awareness one has of oneself in thinking or in acting volitionally, namely in speaking, and accordingly "I" refers to the speaker, who is the object of this "direct" awareness of herself in speaking. It might be proposed, with Frege, that "I" expresses *different* senses or modes of presentation on different occasions. That proposal becomes implausible, however, when we look closely at the mode of awareness one normally has of oneself in thinking or, indeed, speaking.[25] Or it might be proposed, with inspiration from Wittgenstein perhaps, that "I" is not used to express *any* mode of self-presentation, or even to *refer* to oneself. This implausible idea loses its clout as we look not only at the phenomenology of self-awareness but at the remarks of Wittgenstein that prompted it.[26]

What of the cogito?

On this intentional semantics, the sentence "I exist" is *analytically true* – that is, true in any context of utterance by virtue of its semantic character. (The details are the same as in Kaplan's semantics, except there is an additional level of meaning involved, namely the propositional sense, or thought content, "I exist." This sense is not idle, but its role emerges properly in the cognate semantics of thinking "I exist," as opposed to saying "I exist.") However, this semantics does not in itself properly ground the inference "I am thinking, therefore I exist." To be sure, the semantics might *presuppose* that in speaking one is thinking and therefore exists. Specifically, it might be assumed in setting up the semantics that: (a) for any sentence "p" and any context of utterance, the speaker *uttering* "p" is also in that context *thinking* "p," where the content of his so thinking – the proposition "p" – is the sense expressed by his uttering "p" in that context (in the way language expresses thought); and (b) that act of thinking requires the existence of its subject, who is the speaker and thus the referent of "I" in that context. Whence, for any context of uttering "I am thinking, therefore I exist," "I am thinking" is true in that context because the speaker is thinking "I am thinking," and "I exist" is true in that context because the referent of "I" is the speaker, who is assumed to exist in that context. But what is the relation between the speaker's so *thinking* and his existence? That is the question of the cogito, and that relation is not properly articulated among the system of semantical relations codified in the semantics itself. Rather, that relation is articulated in the phenomenology, or "semantics," of thought itself – to which we now turn.

Hintikka suggested that the logic of speech acts should carry over to the logic of thought acts. How might this go? Let us sketch a *semantics* (or formal *pragmatics*) for acts of *thought* (as opposed to speech), formalizing features of the intentionality of thought. We distinguish for each *thought type* four levels of intentional structure (or meaning) for *thought acts* of that type: their intentional character, content, object, and extension. Take the thought type, thinking "I am tall." The *intentional character* of this thought type – if you will, its role or rule of use – specifies for each context (or occasion) of so thinking the content and object (and hence extension) of the act of so thinking in that context. The *content* of this thought act (its "sense") is the proposition "I am tall": we assume that thought types are individuated by thought contents (they are one-to-one). The *object* of this thought act (its "intentional object" – if you will, its "intention" or "intension") is the state of affairs that consists in the subject of this thought act having the property of being tall. And the extension of this thought act is its truth value, where it is true if and only if its "intended" state of affairs is actual.

Now, an act of thinking "I am tall" includes as a constituent a presentation of "I," and correlatively the mode of presentation "I" is a constituent of the thought type "I am tall." The *intentional character* of the mode of presentation "I" specifies for any context of presentation the content and object (and extension) of the presentation of "I" in that context. The *content* of the presentation is the first-person mode of presentation "I," the *object* of the presentation is the subject who is having it, and the *extension* of the presentation is (also) that subject. Importantly, we assume there is an essentially indexical content "I," whose intentional force – as specified by the cognate intentional character – is, where entertained in a particular presentation in a given context, to prescribe the subject of the presentation so entertaining it. This mode of presentation is precisely the phenomenological character of the inner awareness one has of one's own experience, as appraised earlier. And in this mode of presentation one is "directly" aware of oneself insofar as the content "I" directly indicates the subject who is being presented: the content "I" in a presentation of this type prescribes the subject of that presentation solely in virtue of the subject's "performing" that presentation.[27]

And the cogito?

According to this semantics of thought, the proposition "I exist" is *analytically* (*semantically or pragmatically*) *true*: it is true whenever anyone thinks it, by virtue of the intentional character of that thought type. For in any context in which someone thinks "I exist," the content "I" prescribes the person who is so thinking and that person exists in that context, and hence in that context "I exist" is true. Now, this pragmatic analyticity is not the same thing as the indubitability Descartes sought (namely, the self-evidence of thinking itself, which comes with inner awareness). But if in the act of consciously thinking "I exist" the subject intuitively grasps the intentional (or logical) force of the content "I exist," then in that way the proposition "I exist" is indubitable for him in so thinking. And that is what the special cogito principle proclaims. On this semantic account, however, the indubitability of one's existence in thinking "I exist" comes not from one's inner awareness of one's so thinking but from one's semantic intuition of the intentional force of the proposition "I exist" in one's thinking it. The act of thinking "I exist" illustrates that force for the subject, by performing it (*pace* Hintikka).

Furthermore, in any context in which someone thinks "*p*" (for any proposition "*p*"), that person exists in that context, and hence the proposition "I exist" is true of her in that context; or, rather, it would be true if she were thinking it in that context. In that sense the proposition "I exist" is analytically true: in any context of thinking, it would be true if thought in that context. Again, this pragmatic analyticity is not the indubitability Descartes sought for "I exist" (in the subject's inner awareness of her thinking "*p*"). However, insofar as the subject intuitively grasps the intentional force of the content "I exist" while being aware of thinking "*p*," the proposition "I exist" is indubitable for her in thinking "*p*." And that is the claim of the general cogito principle. Again, on the present semantic account, this certainty of one's existence in thinking comes not from one's inner awareness of one's thinking, but from one's semantic intuition of the intentional force of the proposition "I exist" in the context of one's thinking. Similarly, the inferential proposition "I am thinking, therefore I exist" is true in any context in which someone is thinking (no matter what she is thinking), or would be true if the subject were thinking it (or performing the inference) in that context (as opposed to, or in addition to, thinking what she is thinking in that context). And insofar as the subject intuitively grasps its intentional force, it is indubitable – which is the claim of the general cogito inference principle.

It should now be clear that we need to distinguish two sources of evidence that might ground the cogito, and they are appropriate for different cogito claims. On the one hand, there is the evidence in inner awareness of thought; on the other hand, there is the evidence in one's semantic intuition of the intentional force of the proposition "I exist" or "I am thinking, therefore I exist." These two types of evidence give different grounds for being unable to doubt one's existence when thinking. Thus, the evidence of inner awareness most relevantly grounds the following form of the general cogito principle.

> Whenever I am thinking, I am certain that I am so thinking and that I exist.

The evidence of semantic intuition most relevantly grounds the following form of the special cogito principle.

> Whenever I am thinking "I exist," the proposition "I exist" is indubitable (for me in so thinking).

By contrast, the general cogito inference principle holds:

> The inference "I am thinking, therefore I exist" is indubitable (for me in so performing it).

And this claim is most relevantly grounded in semantic intuition, although that semantic intuition is itself grounded in reflection on inner awareness.

Thus, the semantics (or logic) of the cogito is grounded ultimately in the phenomenology and epistemology of inner awareness: the verification, hence certainty, of the proposition "I exist" – or "I am thinking, therefore I exist" – occurs in the act of thinking the proposition and in virtue of the evidence in one's inner awareness of so thinking. However, a different kind of evidence, hence certainty, comes with an act of reflection on thought and content, wherein one comes to "intuitively grasp" the semantics of the propositions "I exist" and "I am thinking, therefore I exist" – namely, their intentional character and thus their pragmatic analyticity, or self-verifying character. Notice that we are dealing with a *second-order* thought in the reflective, semantic judgment that the proposition "I exist" – or "I am thinking, therefore I exist" – is self-verifying (in thinking the proposition).[28] Not only is the *evidence* in the second-order judgment (thought) different from that in the first-order thought, but the *content* is different. Accordingly, the special cogito principle is properly a principle about the proposition "I exist" (in Descartes's slightly edited words: the proposition "I exist" is necessarily true whenever I think it). Whereas the general cogito principle is properly a principle about one's certainty of one's thought and existence in thinking.

The Ontology of the Cogito

What is the nature of the *subject* – the "I" or ego – that is revealed in my inner awareness of my thinking? Am I a purely mental substance (as Descartes argued in the second and sixth of the *Meditations*), or am I a psycho-socio-physical human being (as we assume in everyday life)? And what is the nature of my *thinking* itself which is revealed in inner awareness? Is thinking a purely mental process, a mode of the attribute of thought in a mental substance (as Descartes assumed in the second of the *Meditations*)? Or is it a brain process, or a functional, perhaps computational process executed in a brain? These are the traditional issues of the mind-body problem, central to the ontology of the cogito. We shall not resolve them here (surprise!); I would only point out that these issues are undecided by the phenomenology, epistemology, and logic or semantics of the cogito.

Inner awareness does not reveal the nature of thought or subject beyond what is specified in the form of inner awareness (as explicated earlier, "phenomenally in this experience I am thinking 'p' "). In my awareness of my thinking "Descartes was smart," it clearly *seems* to me (with a certain type of evidence) that I am so thinking and that I exist in so thinking. And it seems to me that my experience of thinking is just that, an *experience* of so thinking, and that I am a *being that so thinks*. However, it does not *seem* to me, in inner awareness, that my thinking is a brain process. Nor does it seem to me in inner awareness that I am a neural system in a human body – or, for that matter, a self defined by interaction of the psychic structures called ego, superego, and id. Yet all this and more may well be true of me and my thinking, so far as the evidence of inner awareness attests. Indeed, it does not *seem* to me in inner awareness that I am a purely mental substance. Only if Descartes was right in his doctrine of "clear and distinct ideas" and its application to the ideas "I" and "I am thinking," only then does the evidence of inner awareness – together with these doctrines – dictate that I am a purely mental substance and my thinking a purely mental attribute (or mode) of that substance. But all that Cartesian machinery outruns the force of the cogito as explicated here. Still, the question remains whether there is a substantial "I" in the limited sense of an *entity* that does the thinking disclosed in inner awareness.

When Descartes proclaimed the cogito inference indubitable, he was accused of presupposing the existence of a substance that has the attribute of thinking (in various modes). Does the semantics of the cogito,

as detailed here, presuppose something similar? According to that semantics, the proposition "I exist" is analytically true, and the sentence
"I exist" self-verifying, just because the relevant context is defined as including a subject or speaker: in defining the semantical relationships, it
is simply presupposed that an utterance of "I exist" is performed by a
speaker, or an act of thinking "I exist" is performed by a subject. Again,
the inference "I am thinking, therefore I exist" is valid (and sound)
just insofar as the semantics presupposes an ontology of subject and
thought. So the phenomenology of inner awareness presents one with
a subject of experience (in the form of awareness "I am thinking . . . "),
and the semantics of thought codifies that deliverance of phenomenology. But we need to ask whether the ontology so presupposed – a subject
for every thinking – is a matter of metaphysical as opposed to logico-
semantic necessity. Diverse thinkers – Hume, Sartre, and early Husserl,
not to mention the Buddha – have held that there is no substantial
self, that there is only the unified stream of consciousness. Without
here evaluating these claims, I would observe that the ultimate nature
of the "I" outruns the phenomenology, epistemology, and semantics of
the cogito: the cogito principles explicated here can accommodate a variety of metaphysical theories about the ultimate nature of thought and
subject.

The cogito principles leave undecided not only the ontological type of
the "I" and the "I think" (mental, neural, or what have you), but also the
ontology of the relations between the thought or subject and the context,
or *environment*, of thinking – if you will, the ecology of thought. The environment includes the physical, social, and even linguistic environment;
the relations to environment are sometimes called contextual relations.
Tyler Burge has argued that thoughts (acts of thinking) are individuated,
in psychological type and content, partly by their contextual relations (including relations to language used to express them).[29] Thought and content, he argues, are not "individualistic" – individuated by factors solely
within the bounds of the individual thinker – but rather "relational" and
especially social insofar as they are individuated by environment. Addressing Cartesian intuitions, he has allowed that we have access to our own
thoughts such that we cannot be wrong about what we are thinking and
yet we can fail to know how those thoughts are individuated in relation to
the environment. I believe there is something right and something wrong
about Burge's arguments, as I have explained elsewhere.[30] Suffice it to
say here that the phenomenology of the cogito leaves open a good deal of
the ontology, including the ecology, of thought.[31] Thus, inner awareness

reveals the phenomenological structure or content of an act of thinking but leaves open the ways in which the thought itself depends on its
environment – for instance, on how the subject's perceptions are caused
(where thinking about something one sees), how the thought is produced
in the subject's brain (or depends on what is produced therein), how the
subject's culture and language developed so as to shape the content of
the thought, and so on.

The Cogito Postmodern

From our vantage approaching the new millennium, I submit, the cogito
resolves into a number of principles articulating, as detailed earlier:

1. The phenomenology of inner awareness of one's experience and
   of oneself.
2. The epistemology of the evidence of inner awareness and the
   reflective-introspective judgments it warrants.
3. The logic/semantics/pragmatics of the relevant speech acts.
4. The cognate "semantics" of the relevant thought acts.

These principles leave open much of the ultimate ontology of thought
and subject.

These principles define (or refine) the Cartesian intuitions of the original *cogito ergo sum*. Yet their motivations are non-Cartesian (they are not
addressed to the particular epistemic and cultural problems of Descartes's
day), and they do not entail the rationalist, foundationalist, or dualist
doctrines of Descartes's wider philosophy. Thus, they define the postmodern cogito. (If they are not congenial to some who seek after "postmodernism," so much the worse for "isms.")

Notes

1. In *The Philosophical Writings of Descartes*, vols. I and II, trans. John Cottingham,
   Robert Stoothoff, and Dugald Murdoch (Cambridge: Cambridge University
   Press, 1984–85), hereafter cited as CSM I or II. In the margins of the CSM
   volumes are page references to the standard edition of Descartes's works: *Oeu-
   vres de Descartes*, ed. Ch. Adam and P. Tannery, rev. ed. (Paris: Vrin/C.N.R.S.,
   1964–76). Page references to CSM are followed, after a slash, by the page references to the AT edition. Most of the passages quoted here are also found
   in the one-volume abridgment of the two CSM volumes: *Descartes: Selected
   Philosophical Writings*, trans. John Cottingham, Robert Stoothoff, and Dugald
   Murdoch (Cambridge: Cambridge University Press, 1988).

2. Emotions belong in this list. In *The Passions of the Soul* (1649) (CSM I, pp. 325–404/AT, pp. 323–488), Descartes classified thoughts as either "actions" or "passions" of the soul (cf. p. 338/AT, p. 349), meaning active or passive activities of the mind, and he went on to enumerate various mental activities including many emotions.

3. In the *Sixth Replies*, Descartes explicitly extends one's certainty to one's thinking as well as one's existence (CSM II, p. 285/AT, p. 422); the passage is quoted and discussed here below. As Wilson (1978: 53ff.) presents the "naive interpretation" of the cogito as inference, the indubitability of "I think" is "a sort of *datum.*"

4. In fact, Descartes had a rich notion of intuition, connected with his doctrine of "perceiving" clear and distinct ideas. Logical intuition might then be cast either as an intuition of a complex idea, namely an inferential proposition, or as an intuition of a temporal sequence of ideas had in entertaining a series of propositions while performing a deduction. This form of intuition would be different from the form of "perception" one has of one's own experiences, to be discussed later. In any event, a sufficiently general doctrine of intuition would allow Descartes to assimilate the intuition that "I am thinking" and the intuition that "I am thinking, therefore I exist." I owe this observation about the cogito to Clotilde Calabi, reflecting on the roles of attention and memory in deduction as addressed by Descartes in *Rules for the Direction of the Mind* (1628). In particular, see Rule Eleven (CSM I, p. 37/AT, p. 407), concerning "intuiting a number of simple propositions... in a continuous... train of thought," and Rule Sixteen (CSM I, p. 66/AT, p. 454), concerning the use of "concise symbols" so that "[i]t will be impossible for our memory to go wrong."

5. See Sartre 1966: introd., sec. III.

6. The content quotes around "*p*" in "I am thinking '*p*' " entail that expressions in "*p*" occur in nonreferential or opaque position, citing contents rather than objects of thought. The basic issues – concerning intentional acts *de dicto* and *de re* and their attributions – are treated at length in Smith and McIntyre 1982: chaps. 1, 7, 8. Indexical expressions in "*p*" raise further issues, some of which are indicated later.

7. See Smith 1989: chap. 2, or, for a shorter version of the analysis, Smith 1986. I was delighted to find the very words "inner awareness of one's thought and existence" in Descartes's own explication of the cogito – and disappointed that my own awareness of those words in Descartes occurred after the publication of said book.

8. Curley 1978: 179. Curley does a marvelous job of tracing the notion of consciousness as awareness of thinking through Descartes's exchanges with various critics. It is abundantly clear that this notion is explicit in Descartes's thinking, although he vacillates on whether the awareness we have of our own thinking is a distinct act of reflecting on thinking, that is, a separate act of introspection.

9. See Smith 1986, 1989.

10. About the phenomenal quality of an experience little can be said: it must be shown rather than said. Or, better, it must be *experienced*, and so indicated

from the first-person point of view (through a reflexive character), as opposed to *described*, for the hearer and hence indicated from the hearer's
third-person, or other-person, point of view. This is why "qualia" have seemed
so mysterious in discussions in philosophy of mind: the attempt to theo-
rize about them takes place in language, which tries to describe them in
"objective," "theoretical" terms. Importantly, qualia are not restricted to
sensory "feels" in perceptual experiences. That restriction, typical in the
literature, is a hangover from the British empiricist days of "sense-data."
On the contrary, every conscious experience – even thinking "I exist" –
has a characteristic phenomenal quality: in the quotably phenomenological
jargon of Nagel (1987), "what it is like" to have that type of experience,
a quality that is to be described from the "subjective," first-person point
of view.

11. The temporal structure of experience – part of its phenomenological structure – was ably chronicled in Husserl's famous analysis of the structure of
"internal time." See Husserl 1966; Husserl's views are reconstructed in Miller
1984.

12. Descartes sometimes treated inner awareness as distinct from introspection,
or reflective thought about thought, and sometimes he treated them as the
same, in part to avoid problems of infinite regress (if thinking requires thinking about thinking, a regress appears to ensue). The vacillation is clearly detailed in Curley 1978: 177–87. Such problems of regress are raised by Sartre,
for instance, and are detailed in Smith 1989: chap. 2.

13. See Husserl 1960/1931: §§9, 18. For a detailed reconstruction of Husserl's
phenomenology of our consciousness of time and the temporal character of
our experience, see Miller 1984.

14. See Husserl 1960/1931: §§5–6.

15. See Ryle 1965. Ryle himself used terms other than "incorrigible." Ryle aptly
distinguished the "awareness" that makes a mental state or activity conscious from the attentive act of "introspection." According to Ryle's gloss
of "the official doctrine" of Descartes and followers, both immediate awareness and introspection give one a "privileged access" to one's own mental
states and operations, and both are "immune from illusion, confusion or
doubt" (p. 14). A careful critique of Ryle's gloss of Descartes in regard to
incorrigibility is found in Curley 1978: 171ff. Suffice it to say that Descartes
was not a "Cartesian" as Ryle reconstructed Cartesian theory. Nor was
Husserl.

16. Cf. Husserl 1960/1931: §§9 and preceding.

17. Hintikka 1962a. The essay is reprinted in Sesonske and Fleming 1965: 50–
76. Page references in text are to this anthology. Hintikka's essay is a carefully
crafted piece that repays line-by-line reading, even where one would differ
(as I shall). Further essays on the cogito as inference or performance are
found in Curley 1978 and Wilson 1978.

18. A detailed survey of the issues of free logics is Bencivenga 1986.

19. See Katz 1987. Katz raises a host of important issues about the boundaries
and relations between logic and language. I raise still others about the
boundaries and relations between logic, language or linguistic theory, and

phenomenology. However, I am pressing the phenomenological and logical, or semantic/pragmatic, sides of the cogito inference.

20. In this century it has proved useful – *extremely* useful – for logic to focus on sentences, in formally defined languages, studying their structure and interpretation and relations of inference or entailment between sentences. In earlier centuries, however, logic often focused on judgments, and indeed Frege's semantics took sentences to express "thoughts." Today we need to extend the logic of sentences to a formally precise logic of not only speech acts but also thought acts and other forms of intentional mental activity such as volitions and desires. Of course, we already have many studies in the logic of *sentences* about such activities, but I am calling for studies in the logic of these activities themselves. Such a logic (or logics) would be part of phenomenology and cognitive science, suitably defined.

21. The more relevant sources for my purposes here are Hintikka 1962b, 1975; Montague 1972; and Kaplan 1977, 1979. Relations between the logic or semantics of sentences ascribing intentional attitudes and the theory of the intentionality of the attitudes are appraised in Smith and McIntyre 1982.

What are the boundaries of the disciplines we call logic, semantics, and pragmatics? Hintikka's appraisal of the cogito called for widening the boundaries of logic to include speech-act theory and even mental-act theory (thus phenomenology). Katz's (1987) appraisal of the cogito called for widening the boundaries of logic to include semantics, in particular, the theory of sense and entailment. Kaplan's appraisal of indexical language widens the boundaries of logic (along lines charted by Richard Montague and Dana Scott) to include the context-sensitive language of indexicals. Logic had already widened, with the logic-*cum*-semantics of modal expressions, to include formal semantics, especially possible-worlds semantics. Pragmatics is traditionally defined as that part of linguistic theory that concerns the *use* of language, including the influence of the context of use. Kaplan's and Montague's systems carry the bounds of logic through formal semantics into formal pragmatics. Let us count as part of the logic of the cogito all the issues indicated, broadening logic to include relevant issues of the form, meaning, and use of language (normally assigned to the domains of syntax, semantics, and pragmatics) and parallel issues of the form, content, and performance of mental activities (normally assigned to phenomenology and epistemology). If it seems strange today to think of logic as concerning mental activities, we need only recall that inference is a movement in thought as well as language, and in earlier centuries logic was primarily concerned with judgments rather than sentences. Today, after a long and fruitful semantic ascent to the realm of sentences and their "logic," we can begin to return to the realm of thought or judgments and their "logic." Here I speak of the "semantics" of the cogito – in language or in thought – because the issues addressed concern the meaning of the sentence "I exist" as used in a speech act or the proposition "I exist" as used in a mental act. If you will, we are here addressing issues of pragmatics that fall within semantics, although other aspects of use may not.

22. As is well known, Russell's semantics posited only one level of meaning, while Frege's semantics posited the two levels of sense and referent. Take the simple sentence "I am tall." In Russell's system: "I" means (or refers to) the individual *a* who utters it, "tall" means the property of being tall, and "I am tall" means the fact or "proposition" that *a* is tall – in effect, the state of affairs consisting of the individual *a* and the property of being tall joined by instantiation. By contrast, Frege's semantics posited two levels of meaning: the "sense" and the "referent" of an expression are distinct, the sense determines the referent, and the sense embodies the "mode of presentation" of the referent. Thus, in Frege's system: "I" expresses a sense that determines the individual *a* who utters it, "tall" expresses a predicative sense that determines the "concept" (in effect, the property) of being tall, and "I am tall" expresses the "thought" that is formed by combining the sense of "I" and the sense of "tall" and which determines the truth-value True (as opposed to the state of affairs that *a* is tall). For Frege, the indexical "I" expresses on each occasion of utterance an appropriate descriptive sense (e.g., that of "the tallest Irvine philosopher") that determines the individual who utters "I" on that occasion. We cannot here go into the many questions raised in the interpretation of Frege's and Russell's systems, but the preceding gloss can serve as a foil for Kaplan's system.

23. Let us assume that existence is a property of individuals. There àre, of course, many problematic features of existence. Is it a bona fide property, or a different category altogether, as suggested by Kant's dictum, "Existence is not a predicate"? Is it a property of properties rather than individuals, à la Frege? Is it paralleled by the property of nonexistence, à la Meinong? Is it an indexical property, namely, that of occurring in "this world," à la David Lewis? Is it not a property (kind, quality, etc.), but a process, in our case – as Heidegger stressed – the process of Being carried out by human beings? In order to assess the cogito fully, we would ultimately have to develop an ontology of the property of existence and its instantiation in thinkers and speakers. But for our purposes here let us adopt a minimal assumption that existence is a property of individuals.

24. An "intensional" logic is usually characterized as a system in which substitutivity of identity fails, for example, in contexts like "Necessarily __" or "s believes that __," and an "intensional" semantics is one which appeals to "intensional" entities, for example, properties or propositions (whatever the latter are taken to be). The present "intentional" semantics invokes "intentional" contents of intentional activities like belief, thought, perception, or intention. Cf. Smith 1981. The semantical scheme sketched here differs in some ways from what I proposed there. (1) Here I make the character, or rule of use, explicit as a level of meaning; there I left it implicit in the semantical system. (2) Here I use the term "proposition" for the sense (= content of thought) expressed by a sentence; there I capitulated to Kaplan's neo-Russellian use of "singular proposition" for singular states of affairs, which would be misleading in this essay, given Descartes's use of "proposition" (in translation in CSM), not to mention the wider phenomenological traditions that differ from the Russellian use. (3) Here I sketch a parallel semantics

of thought; there I did not. (4) Here I skirt the use of "possible worlds" in the semantics, which would take us too far afield from the Cartesian issues; there I gave them an explicit place in the semantics.

25. The first-person mode of awareness "I" is studied in detail in Smith 1989: chap. 2. Frege (1967) proposed that "I" expresses a different descriptive sense on each occasion of utterance. Perry (1979) argued effectively against Frege's view. My account of the word "I" and the sense, content, or mode of awareness "I" would join Perry's suit against Frege's claim.

26. Anscombe (1975) has argued, with inspiration from Wittgenstein, that "I" is not a referring expression, from which one might infer that it does not serve to express any mode of presentation of oneself. She argued basically that if "I" refers, it must refer to a Cartesian mental substance, which is objectionable, hence "I" is not a referring expression. Wittgenstein's own remarks were more subtle. Wittgenstein (1958: §410) said "I" is not the name of a person (cf. §§398–428 on "I" and pain). In "The Blue Book" Wittgenstein (1965) had argued, in more detail, that "I" is used in two ways: the use as object in, for example, "I have grown six inches"; and the use as subject in, for example, "I see so-and-so," "I think it will rain," "I have toothache," and so on (pp. 66–70). Wittgenstein referred explicitly to the cogito, writing:

> We feel that in the cases in which "I" is used as subject, we don't use it because we recognize a particular person by his bodily characteristics; and this creates the illusion that we use this word to refer to something bodiless, which, however, has its seat in our body. In fact *this* seems to be the real ego, the one of which it was said, "Cogito, ergo sum." (p. 69)

More recent accounts of indexicals would grant that "I" is not used as a name and does not rely on the speaker's recognizing his bodily traits. A cognate issue is the logic of sentences like "I believe that I have grown six inches," in which "I" is used once "as subject" and once again "as object." Castañeda (1966) stressed the special logic of "he himself" in "John believes that he himself is a Wittgensteinian," which ascribes a belief the subject would report by saying, "I am a Wittgensteinian." Perry (1979) followed Castañeda in arguing the irreducibility of indexicals. Although the issues are complex and would take us far afield, I believe Wittgenstein's remarks on the uses of "I" are not inconsonant with an account of "I" geared to the phenomenology of self-awareness: see the account sketched in the text, and consider a thought whose phenomenological structure is articulated by "Consciously I now am thinking that I have grown six inches," which involves awareness of oneself as subject ("I am thinking . . .") and as object ("I have grown . . ."). The reader who thinks Wittgenstein is wholeheartedly antiphenomenological should study Hintikka and Hintikka 1986.

27. Again, this mode of presentation "I" is studied in Smith 1989: chap. 2. To be exact, the distinction is there drawn between being presented oneself as in thinking "I am tall" and being aware of oneself in inner awareness proper, as "In this very experience I am thinking 'Kareem is tall.'"

28. The term "second-order" is used in different ways. Thoughts about thoughts are called second-order thoughts, propositions (= thought contents) about

propositions are called second-order propositions, and quantifiers (linguistic items) are called second-order quantifiers if they range over propositions, whence sentences about propositions might be called second-order if they involve second-order quantifiers.

29. Cf. Burge 1979. Burge (1988) later seeks to accommodate some Cartesian intuitions.
30. See Smith 1990.
31. This "leaving-open" is precisely the role of what Husserl called the "horizon" of an experience. Cf. Husserl 1960/1931: §§18–20 and Smith and McIntyre 1982: chap. 2.

References

Anscombe, G. E. M. 1975. "The First Person." In S. Guttenplan, ed., *Mind and Language*, pp. 45–65. Oxford: Clarendon Press.

Bencivenga, Ermanno. 1986. "Free Logics." In D. Gabbay and F. Guenther, eds., *Handbook of Philosophical Logic*, 3:373–426. Dordrecht: D. Reidel.

Burge, Tyler. 1979. "Individualism and the Mental." In Peter A. French, Theodore E. Uehling, and Howard Wettstein, eds., *Midwest Studies in Philosophy*, IV: *Studies in Metaphysics*, pp. 73–121. Minneapolis: University of Minnesota Press.

1988. "Cartesian Error and the Objectivity of Knowledge." In Robert H. Grimm and Daniel E. Merrill, eds., *Contents of Thought*, pp. 62–76. Tucson: University of Arizona Press.

Castañeda, Hector-Neri. 1966. "'He': A Study in the Logic of Self-Consciousness." *Ratio* 8: 130–57.

Curley, E. M. 1978. *Descartes against the Skeptics*. Cambridge, Mass.: Harvard University Press.

Descartes, René. 1984–85. *The Philosophical Writings of Descartes*. Vols. I and II. Translated by John Cottingham, Robert Stoothoff, and Dugald Murdoch. Cambridge: Cambridge University Press.

1988. *Descartes: Selected Philosophical Writings*. Translated by John Cottingham, Robert Stoothoff, and Dugald Murdoch. Cambridge: Cambridge University Press.

Frege, Gottlob. 1967. "The Thought: A Logical Inquiry." In P. F. Strawson, ed., *Philosophical Logic*. Oxford: Oxford University Press.

Hintikka, Jaakko. 1962a. "*Cogito, Ergo Sum*: Inference or Performance?" *Philosophical Review* 71: 3–32. Reprinted in Alexander Sesonske and Noel Fleming, eds., *Meta-Meditations: Studies in Descartes*, pp. 50–76. Belmont, Calif.: Wadsworth, 1965.

1962b. *Knowledge and Belief*. Ithaca: Cornell University Press.

1975. *The Intentions of Intentionality*. Dordrecht: D. Reidel.

Hintikka, Merrill B., and Jaakko Hintikka. 1986. *Investigating Wittgenstein*. Oxford: Basil Blackwell.

Husserl, Edmund. 1960. *Cartesian Meditations*. Translated by Dorion Cairns. The Hague: Martinus Nijhoff. German original, 1931.

1966. *The Phenomenology of Internal Time-Consciousness*. Translated by James S. Churchill. Bloomington: Indiana University Press. German original, from lectures of 1905–10.

Kaplan, David. 1977. *Demonstratives*. University of California, Los Angeles. Typescript. Revised and updated version appeared as "The Logic of Demonstratives," in Joseph Almog, John Perry, and Howard Wettstein, eds., *Themes from Kaplan*. Oxford: Oxford University Press, 1989.

1979. "On the Logic of Demonstratives." *Journal of Philosophical Logic* 8: 81–98.

Katz, Jerrold J. 1987. *Cogitations*. Oxford: Oxford University Press.

Miller, Izchak. 1984. *Husserl, Perception and Temporal Awareness*. Cambridge, Mass.: MIT Press.

Montague, Richard. 1972. "Pragmatics and Intensional Logic." In Donald Davidson and Gilbert Harman, eds., *Semantics of Natural Language*. Dordrecht: D. Reidel.

Nagel, Thomas. 1987. *The View from Nowhere*. Oxford: Oxford University Press.

Perry, John. 1979. "The Problem of the Essential Indexical." *Nous* 103: 3–21.

Ryle, Gilbert. 1965. *The Concept of Mind*. New York: Barnes and Noble.

Sartre, Jean-Paul. 1966. *Being and Nothingness*. Translated by Hazel Barnes. New York: Washington Square Press.

Smith, David Woodruff. 1981. "Indexical Sense and Reference." *Synthese* 49: 101–27.

1986. "The Structure of (Self-)Consciousness." *Topoi* 5(2): 149–56.

1989. *The Circle of Acquaintance*. Dordrecht: Kluwer Academic Publishers.

1990. "Thoughts." *Philosophical Papers* 19 (3): 163–89.

Smith, David Woodruff, and Ronald McIntyre. 1982. *Husserl and Intentionality*. Dordrecht: D. Reidel.

Wilson, Margaret Dauler. 1978. *Descartes*. London: Routledge and Kegan Paul.

Wittgenstein, Ludwig. 1958. *Philosophical Investigations*. Translated by G. E. M. Anscombe. New York: Macmillan.

1965. *The Blue and Brown Books*. New York: Harper Torchbooks.

3

Return to Consciousness

Abstract. What makes a mental state conscious, according to the classical view (in Locke et al.), is a certain *self-consciousness*, or (as I prefer to put it) an *inner awareness* of the state. What is the *form* of that inner awareness? This is a difficult question, as we see by studying neoclassical models in Brentano, Husserl, and others. In recent philosophy of mind it has been proposed that this awareness of our experience consists in a higher-order monitoring. Yet there are problems with all higher-order theories of consciousness, as Brentano well observed. Here I pursue and partly revise my own earlier analysis of inner awareness as a "modal" character of mental acts. On that analysis, inner awareness is an integral part of an act of consciousness; it is not a higher-order act of any type (such as observing one's current thought or perception). On a particular account, this inner awareness may itself be grounded in the temporal flow of consciousness (extending Husserl's analysis of time consciousness). Yet, in the end, we should allow that lower forms of consciousness do not include the form of inner awareness typical of everyday human experience. Consciousness does not, then, reduce to inner awareness; instead we need a systematic classification of types and levels of consciousness, and we sketch the beginning of such a classification.

Segue. In "The Cogito circa A.D. 2000" we appraised the core elements of the Cartesian account of how consciousness is *eo ipso* a consciousness of its object and an awareness of itself. Here we pursue the form of that inner awareness, and we consider limitations on the classical view of consciousness as self-consciousness. This exploration is purely phenomenological and focused indeed on the most "inner" aspect of conscious experience. Later chapters, however, lead from phenomenology into ontology, expanding the horizons of our own consciousness.

I thank Amie Thomasson and Jason Ford for comments on the penultimate version of this chapter.

Some years ago (in *The Circle of Acquaintance* [1989] and, tersely, in "The Structure of (Self-)Consciousness" [1986]), I proposed a model of the phenomenological structure that makes a mental act or state *conscious*: what I called an *inner awareness* of the act taken as a *modal* structure distinguished from any higher-order reflection or judgment or recollection or perception of the original act. Since that time, consciousness has come center stage in philosophy of mind *cum* cognitive science. Philosophy has returned to consciousness, after the dark ages of behavioristic theory (from Skinner to Ryle to Dennett). And now I return to the story of consciousness, seeking to refine a neoclassical analysis of consciousness.

Here I hope to deepen and widen – and, in some ways, revise – my earlier modal analysis of inner awareness by contrasting it with alternative views drawn from both phenomenology and cognitive science. The discussion addresses observations of blindsight and theories of higher-order monitoring, in contemporary philosophy of mind allied with cognitive science; the analysis of inner consciousness and time consciousness, in Brentano's psychology and Husserl's phenomenology; and my own earlier account of modality in intentionality. Finally, we consider certain limits on inner awareness.

In this way I hope to bring out the uniquely *modal* form of inner awareness in consciousness.

Consciousness in Historical Contexts

The theory of consciousness, like most things modern, took root with Descartes.

In his *Meditations on First Philosophy* (1641), in the Replies to Objections, Descartes held that each thought includes a certain "internal thought" (*cognitione interna*) of itself.[1] This form of cognition is *internal* to the original thought: it cannot be a second, reflective thought on pain of infinite regress (entailing thought of thought of . . .). (See my appraisal of Descartes's view in "The Cogito *circa* AD 2000" [1993], reprinted in Chapter 2.) The problem of how to understand this inner cognition, or inner awareness, is the central motif in all that follows in the present essay.

Locke (1694) developed the conception of inner cognition much more explicitly, in the following terms, which set the problem in a rich phenomenological context:

[T]o find wherein *personal identity* consists, we must consider what *Person* stands for; which, I think, is a thinking intelligent Being, that has reason and reflection, and can consider it self as it self, the same thinking thing in different times and places; which it does only by that consciousness, which is inseparable from thinking, and as it seems to me essential to it: It being impossible for any one to perceive, without perceiving, that he does perceive. When we see, hear, smell, taste, feel, meditate, or will any thing, we know that we do so. Thus it is always as to our present Sensations and Perceptions: And by this every one is to himself, that which he calls *self*.... For since consciousness always accompanies thinking, and 'tis that, that makes every one to be, what he calls *self*, and thereby distinguishes himself from all other thinking things, in this alone consists *personal identity, i.e.,* the sameness of a rational Being: And as far as this consciousness can be extended backwards to any past Action or Thought, so far reaches the Identity of that *Person.* (1975/1694: 335)

Here in an early form are just the issues we shall be treating as we proceed, not least as we turn to Brentano and Husserl.

Leibniz (1714) called inner cognition "apperception" or "consciousness," implying a unique form of cognition: "[I]t is important to make a distinction between *perception*, which is the internal state of a monad [or mind] which represents external things, and *apperception*, which is *consciousness*, or the reflective knowledge of that internal state."[2] Notice that Leibniz calls this apperception "reflective knowledge" of the internal state, and here is how he characterizes the mind: "Each monad [or mental substance], together with its own body, makes up a living substance.... A living thing of this kind [with appropriate complexity of organs and feeling, memory, and reason] is called an *animal*, since its monad is called a *soul*. And when that soul is at the level of *reason*, ... we count it as a mind."[3] Leibniz's words are eerily familiar today despite his seemingly quaint notion of "monad."[4]

In classical philosophy and psychology, however, the sharpest and most detailed statement of the problem of consciousness as inner awareness is found in Franz Brentano's *Psychology from an Empirical Standpoint* (1874). And recall that Brentano was well versed in Aristotle, the medieval philosophers, the empiricists, Kant, and nineteenth-century psychology (English as well as German). Reviving the medieval notion of *intentio*, Brentano put forth his famous thesis: every mental act contains an object intentionally within it or (better) is directed toward an object (pp. 88ff.). As we say today, every mental act is *intentional*. But Brentano adjoined to this thesis a second principle: every mental act also includes an "inner consciousness" of itself (pp. 139ff. 160). Accordingly, Brentano argued (in great detail) that every mental act is conscious and that each act is a consciousness of its object and also of itself. Brentano struggled valiantly to articulate this secondary inner consciousness as a dependent, inseparable part of the given act, and not a separate higher-order act directed toward the act. Thus, "without any further complication and multiplication of entities," Brentano held (p. 154), my hearing a sound is, in different respects, both a consciousness of the sound and a consciousness of the hearing. It has been argued accordingly (in Thomasson 2000) that Brentano's position should be read and further developed as a one-level theory, rather than a higher-order theory, of consciousness. Here we have the modern problem of consciousness, which hovers with us today: how to articulate such an inner awareness. We take a closer look at Brentano's analysis shortly.

Edmund Husserl extended his teacher Brentano's analysis of both intentionality and inner consciousness. In *Logical Investigations* (1970/1900–1) Husserl drew (in fine detail for the first time) the fleet of distinctions that characterize *intentionality* (Husserl introduced this term of art): an act of consciousness, experienced or performed by a subject or ego, is directed via a content or meaning toward an object (if such object exists) – where the meaning characterizes the object "as experienced," resting on a background or "horizon" of associated meanings, variously shaped by one's culture and increasingly by the results of empirical science. But Husserl also held, like Brentano, that every act has not only a *primary intentionality* but also a *secondary intentionality*: in the first form of intentionality, the act is directed toward its object; in the second, toward itself. This secondary intentionality Husserl analyzed further, as we consider later, in terms of one's consciousness of the temporal flow of experiences.

Recently, philosophers, psychologists, and neuroscientists working in the tradition of cognitive science have rediscovered the classical range of issues about consciousness, including issues about inner sense, apperception, or secondary intentionality gathered under the banner "higher-order monitoring." A valuable and extensive compendium of writings in this genre is the anthology *The Nature of Consciousness* (Block, Flanagan, and Güzeldere 1997). It is important to stress that the problem of consciousness is *phenomenological* – a word, crystallized in the work of Husserl, that appears with increasing frequency in recent philosophy of mind *cum* cognitive science (as in Block et al.). Yet rarely is this citation of the phenomenological supported by either careful phenomenological analysis or any reference whatsoever to the extant discipline of phenomenology, which has been a rich and going concern at least since Husserl's *Logical Investigations* of 1900, after the psychology of Brentano and also William James (whose work Husserl had also read). Although James's views are sometimes discussed and Brentano is occasionally cited, the rich results of Brentano, Husserl, and subsequent phenomenologists are generally ignored (see Block et al.). An important exception is an emerging European-American tradition that combines phenomenology with cognitive science, well represented in the substantial anthology *Naturalizing Phenomenology: Issues in Contemporary Phenomenology and Cognitive Science* (Petitot et al. 1999).[5] Here I join in that spirit.

It should be tautological that consciousness is what makes the mind-body problem difficult. Yet it was not obvious as Thomas Nagel (1973) laid out this challenge at the beginning of the cognitive-science revolution. Subsequently, John Searle (1992) has argued doggedly for the irreducibly subjective character of consciousness and its first-person ontology. David Chalmers (1996) has argued in close detail that the phenomenal character of consciousness is the "hard" problem for cognitive science today. And Charles Siewert (1998) has argued carefully and at length for a "first-person" approach to consciousness. Because phenomenology is classically defined as the study of consciousness "as we experience it," from the first-person perspective, we need today to return to phenomenology as its own discipline within the wider theory of consciousness that includes much recent work in philosophy of mind and cognitive science.

But what makes consciousness the *really hard* problem for the theory of mind is the problem of inner awareness. Without our own inner awareness of our passing experiences, we might plausibly identify mental states with neural activities whose essence lies in physical properties of neurons and their causal interactions with their physical environs. The character of inner awareness, however, calls out for analysis in its own terms. We need to understand the *form* of inner awareness, which is not the same as the form of its neural implementation, or its causal interactions, or other functional properties of a mental state. And the formal analysis of inner awareness, amid other forms of consciousness, is a task of phenomenology.

From Blindsight to Consciousness

The problem of consciousness has been brought to the fore dramatically by the empirical observation of what is called "blindsight." Indeed, the phenomenon of blindsight puts the character of consciousness itself in sharp relief, and we begin the present discussion there.[6]

Studied in cognitive science in recent years, blindsight is commonly characterized as follows.[7] Some people, having suffered damage to the visual cortex, say they see nothing but are able to answer questions about objects placed before them. They often resist the questions, saying, "I can't see anything" but then go on to say correctly that the object before them is, say, a vertical rather than horizontal line, or perhaps a book rather than a frog. The subject in such experiments, we naturally observe, sees the book before him or her but is not aware of seeing it. Her visual state is one of *blindsight*: she sees, but her seeing is not *conscious*.

In the reported cases, blindsight is not very good at articulating what is before the subject.[8] Still, there is an important aspect of mental structure in the subject's seeing (taking in information visually) *without being aware* of seeing anything. Ideally, by extrapolation, we should note the possibility in principle that for any conscious visual experience there is a corresponding type of vision that lacks the character of consciousness but is otherwise the same in structure. To be sure, we should be cautious in basing theory on thought experiments that far outrun "real" possibilities. And yet the *ideal* structure that distinguishes conscious from unconscious vision is just what phenomenological analysis is after. At least on the classical Husserlian approach (in *Logical Investigations*), phenomenological analysis is a kind of "logical" analysis of ideal mental types or "meaning" structures in familiar forms of experience.[9] So let us recognize a particular aspect of mental structure, the character of consciousness, which we naturally ascribe with the adverb "consciously."

In blindsight, then, one sees but does not *consciously* see. Accordingly, we distinguish two *forms* of visual perception, whose intentional structure (in a simple example) is ascribed as follows:

Consciously I see this frog.
Unconsciously I see this frog.

The first form of perception is conscious; it is conscious sight. The second is not conscious; it is blindsight, or unconscious sight. The difference is that the first has and the second lacks the character of consciousness: the character "consciously" – to give it a name and a basic category to be further examined.

What is this character of *consciousness*, this quality that makes a visual state *conscious*? More generally, what is the character of *consciously* seeing or thinking or wishing or willing: the character that makes a conscious mental act (of whatever type) *conscious*?

First and foremost, the character of consciousness is a *phenomenological form* of a mental act: something we each *experience* in our own first-person conscious experiences – and to which we give voice in first-person phenomenological descriptions like "Consciously I see this frog." In fact, it is only by experiencing conscious sight myself that I can project the character of consciousness into your vision, namely, by empathy.[10] And I so gather that you and others also experience consciousness in conscious vision. Now, in the blindsight experiment we make third-person claims about another's mental state, taking the other's experience in conversing with us to have consciousness but taking her perception to lack the character of consciousness. Here we practice "heterophenomenology," analyzing another's experience. By contrast, in classical phenomenology – "autophenomenology" – we each analyze our own experience, from the first-person perspective, and we assume others have similar forms of experience.[11] Here we see an interaction of method between phenomenology and cognitive science.

Because blindsight is so uncommon, the sentence "I see this frog" in everyday English normally carries a presupposition of consciousness. Distinguishing conscious and unconscious sight, however, allows us to parse off that presupposition for the purpose of our present analysis. We then add the modifier "consciously" to ascribe the character of consciousness, and we use "unconsciously" to mark the lack of that character.

So the phenomenological character "consciously" is, to begin with, the property that distinguishes conscious sight from blindsight. Generally, it is what makes a conscious mental state conscious.

According to the classical analysis, from Descartes and Locke to Brentano and Husserl, consciousness essentially involves an "inner" cognition of one's mental state. Classically, that form of *inner awareness* defines the character of consciousness and makes the state conscious. Indeed, on that analysis, the difference between conscious vision and blindsight is precisely the fact that in conscious vision one is *aware* that one is seeing, whereas in blindsight one is not aware that one is seeing. The experimental observation of blindsight places in sharp relief this property of inner awareness, which is otherwise so familiar that we lose sight of it. Having brought it to salience, we go to work in further analysis.

I choose the term of art "inner awareness," rather than "inner cognition" or "inner perception" or "introspection," for reasons that will unfold. Later we shall ask whether all states of consciousness essentially involve inner awareness, allowing that lower levels of consciousness do not. But in the meantime we focus on the *form* of inner awareness where it is present in familiar forms of consciousness.

Generally, we should allow that there are different types and degrees of consciousness. Freudian psychoanalysis distinguished among the dynamic unconscious (forced by repression), the merely static unconscious (not so forced, arguably with a different character), the preconscious (ideas or emotions awaiting prompting), the strictly conscious, and even the verbalizable conscious.[12] Leaving psychoanalysis to the side, however, there seem to be degrees of consciousness even in human vision. Our everyday experiences of clear and attentive vision include a standard character of "full" consciousness, which arguably includes a form of inner awareness. By contrast, the reported cases of blindsight lack this character. In principle there are gradations of consciousness between these ideals of fully conscious and quite unconscious vision. However, by focusing on ideal cases of conscious and unconscious sight, those which have and lack inner awareness, we turn our attention to the question of the proper analysis of such inner awareness – in paradigm cases where it characterizes conscious visual experience.

The Form of Inner Awareness

What is the *form* of this inner awareness? That is the central question before us. It is a phenomenological question.

Although we shall use the logical or grammatical form of phenomenological descriptions as a tool to express or articulate forms of experience, it is the phenomenological – and ultimately ontological – form of the experiences described that is our concern. This style of phenomenological analysis has been developed elsewhere in greater detail.[13] I stress that forms of language are not the issue at hand, even though we use words to capture (however imperfectly) forms of mind. I interject this remark because there has long been a tendency to conflate forms of language with forms of world – and so, in philosophy of mind, first to conflate intentionality with intensionality (the logic of our language about intentionality), and then to conflate intentionality with computationality (the logic of computer simulations of intentionality, formulated in computer languages). (Perhaps that tendency even defines a certain narrow conception of "analytic" philosophy, of philosophy as logical or linguistic analysis, sometimes assuming we never get beyond our language to reality, mind, politics, or what have you.)

According to my own prior analysis of the structure of consciousness (in Smith 1986, 1989), the form of *inner awareness* defines consciousness per se, or makes a mental act or state conscious, in familiar forms of thinking, perceiving, desiring, willing. Following that analysis, we articulate the *form* of consciousness (staying with the simple example) in the phenomenological description:

Consciously I see this frog.

The character of consciousness, as assumed earlier, finds first-person phenomenological description here in the modifier "consciously."

Importantly, on my analysis (1986, 1989), the character "consciously" is not part of the mental act's intentional relation to an object or its character of object presentation. That is, it is not part of the *mode of presentation* of the object, embodying the content through which the act is directed toward its object, the content "this frog" or perhaps "this poisonous red tree frog." Rather, "consciously" is part of the *modality* of presentation or intentionality: the *way* the perceptual act is executed with its mode of presentation and so its intentional relation to its object. The act's modality (what Husserl called its "thetic character") includes its act character of seeing, as distinct from hearing, thinking, willing. The character "I [see] . . ." is a further character in the act's modality, an *egocentric* character, its being directed from subject toward object. Now, on my proposal, the character "consciously" is a further part of the act's modality, modifying the act's perceptual character so as to make it conscious. The significance of this claim will unfold along our way, distinguishing this modal theory from alternative theories of consciousness.

In closer detail, my prior analysis factored the character "consciously" into two characters ascribed as follows:

Phenomenally in this very experience I see this frog.

The character "phenomenally" embodies the *phenomenal* or *qualitative* character of the experience, commonly now called its quale or (in plural detail) qualia. The character "in this very experience" embodies the *reflexive* character of the experience, its self-reflexive character. The *form* of inner awareness consists in this reflexive character modified by the phenomenal character. Together they define the character "consciously," which modified the whole mental act. As we proceed we shall look more closely at the reflexive character, contrasting it with various forms of high-order monitoring.

We schematize the *form* of the mental act in Figure 3.1, factoring its overall intentional character into several intentional categories. The form of *inner awareness*, on this analysis, is that of "in this very experience" modified by "phenomenally." These characters together define the character

INTENTIONAL CHARACTER

PHENOMENAL	REFLEXIVE	EGOCENTRIC		
				OBJECT-
			ACT-TYPE	PRESENTATION
Phenomenally (in this very experience)	(I)		(see)	(this frog)
MODALITY				*MODE*

FIGURE 3.1. The form of an act of consciousness.

"consciously," on the present analysis, all within what I have called the modality of intentionality or presentation as distinguished from the mode of presentation. Let us be clear that inner awareness need not be infallible or incorrigible. Although Descartes sought demon-proof certainty in the cogito, in my consciousness of my own thinking, we need not follow him in that quest. (This I argued in "The Cogito *circa* AD 2000" [1993], reprinted in Chapter 2.)

Of course, consciousness is present in many types of experience: in seeing a frog, in thinking about Paris, in swinging a bat at a baseball, and so on. Moreover, there are degrees and varieties of consciousness, issuing in variations in the character "consciously." If I am concentrating on this frog and its movement, my perception is more vivid and more salient than is my perception of a frog I am leaping over in a stream. Still less vivid and salient in my experience are my passing feelings of annoyance at my friend's all-too-familiar mannerism. And psychiatry has long posited feelings that are only vaguely conscious or even wholly unconscious and repressed. For simplicity, and present purposes, let us stay with a familiar case such as consciously seeing this frog or consciously thinking that this frog will jump.

Inner Consciousness

The problem of inner awareness comes to maturity, we noted, in Brentano's account of what he called "inner consciousness" (*inneren Bewusstsein*). (See 1995/1874: book 2, chaps. II and III.) Brentano's critique of numerous arguments in prior psychology and philosophy leads smoothly into the back-and-forth of today's debates about the specifics of inner awareness (see Block et al. 1997; Siewert 1998; Thomasson 2000). Moreover, Brentano's own positive account gives us a clear model of what is at stake in the form of inner awareness – even if we reject all the details of his theory.

Brentano distinguished mental phenomena – the domain of psychology – from physical phenomena in the following way:

Every mental phenomenon [or act] is characterized by... the intentional (or mental) inexistence of an object, and what we might call... reference to a content, direction upon an object.... Every mental phenomenon includes something as object within itself, although they do not all do so in the same way. In presentation [*Vorstellung*] something is presented, in judgment something is affirmed or denied, in love loved, in hate hated, in desire desired and so on.

This intentional in-existence is characteristic exclusively of mental phenomena. No physical phenomenon exhibits anything like it. We can, therefore, define mental phenomena by saying that they are those phenomena which contain an object intentionally within themselves. (1995/1874: 88–89)

In other words, every mental act is directed toward an object. But Brentano's conception of this directedness follows a theory – derived from the Scholastics' notion of *intentio* – that the object is intentionally contained "in" the mental act itself. In this way he averted the distinction between content and object, which was pivotal in Husserl's subsequent and much improved theory of intentionality. Brentano's formal analysis of mental acts distinguishes presentation, judgment, love, desire, and so on. For Brentano, we do not merely form a presentation of an object; we also judge that the object exists or that it does not, and we have feelings toward it, as in love or hate, whence we may desire it, and so may will to do something with regard to it. Where presentation presents an object (with certain properties), judgment affirms or denies its existence, feeling (in love, hate, desire, etc.) colors it with emotion, and will initiates action concerning it. Brentano's successors (including Husserl and Ludwig Wittgenstein) advanced the notion of state of affairs (*Sachverhalt*), "that object A has property P" or "that A stands in relation R to B," whence Bertrand Russell put the logician's eye on what he called "propositional attitudes," that is, intentional attitudes whose objects are states of affairs.[14] But Brentano, like Aristotle, focused on objects with properties.

With this structure of directedness behind him, Brentano proceeded to a lengthy study of "inner consciousness" (*inneren Bewusstsein*), tracing the notion to Aristotle's view in *De Anima* that we perceive our sensations, without perceiving such perceptions, averting an infinite regress. Brentano argued roughly as follows. We sometimes "perceive" our own acts of consciousness. This inner perception cannot be "observation," because observation requires full attention, but we usually cannot attend to our own consciousness while going through it (as in feeling angry about something). Nor can inner perception be a second "judgment" to the effect that the act occurs; for if this judgment were itself to be perceived, an infinite regress would ensue. Nor can inner perception be a recollection or memory, which again would lead into an infinite regress. Whether simultaneous with the act or following it, neither a judgment (thought) about it nor a recollection of it – much less a regress of thoughts or memories – is part of our experience. And, of course, inner perception is not a kind of *sensory* perception; it does not consist in a presentation of a sensory object such as a color or a tone, but rather accompanies and is part of such a presentation. Inner consciousness, then, seems for Brentano to take a form unto itself. Brentano considered at length whether this inner consciousness is present in all mental acts, making all mental acts conscious. He concluded that it is because the *intensity* of inner consciousness is the same as that of the act in which it occurs, say, in hearing a tone. This is an odd claim (what is the "intensity" of the inner consciousness as distinct from the intensity of hearing of the tone?). Beneath it seems to lie the presupposition that the inner consciousness is somehow coextensive with the primary consciousness of the tone. The basic question, then, is what *form* this inner consciousness takes, if it is somehow an inseparable part of the act. The details of Brentano's account are instructive.

After criticizing previous theories of the form of inner consciousness (in a tour de force of psychological theories from Aristotle through the English and German philosophers and psychologists of the eighteenth and nineteenth centuries), Brentano summarized his own positive account:

Every mental act is conscious; it includes within it a consciousness of itself. Therefore, every mental act, no matter how simple, has a double object, a primary and a secondary object. The simplest act, for example the act of hearing, has as its primary object the sound, and for its secondary object, itself, the mental phenomenon in which the sound is heard. Consciousness of this secondary object is threefold: it involves a presentation of it, a cognition of it and a feeling toward it. Consequently, every mental act, even the simplest[,] has four different aspects under which it may be considered. It may be considered as a presentation of its primary object, as when the act in which we perceive a sound is considered as an act of hearing; however, it may also be considered as a presentation of itself, as a cognition of itself, and as a feeling toward itself. In addition, in these four respects combined, it is the object of its self-presentation, of its self-cognition, and (so to speak) of its self-feeling. Thus, *without any further complication and multiplication of entities*, not only is the self-presentation presented, the self-cognition is known as well as presented, and the self-feeling is felt as well as known and presented. (1995/1874: 153–54, emphasis added)

A lot of structure is subtly crafted into this analysis of inner consciousness. Recognizing this type of structure guides our investigation, even if we reject the specifics of Brentano's analysis.

Consider an experience in which I hear the tone middle C played on a Steinway piano. The structure of this act of consciousness might be articulated in a phenomenological description something like this, if we assume for the moment Brentano's analysis of four distinct elements in consciousness:

I am conscious in <u>presentation</u>
[1] (<u>externally</u>) of the object Middle C and
[2] (<u>internally</u>) with a <u>feeling</u> of delight in a <u>judgment</u> affirming existence of <u>this consciousness</u> of that object.

Thus, the act of consciousness is directed (externally) toward the primary object, Middle C, and (internally) with feeling and with judgment toward the secondary object, the act itself. Notice that the character of aural sensation is absorbed into the object, the sound middle C (with the special timber of a Steinway grand piano, as it happens). In this simple act of hearing there is a presentation of middle C, and that is all there is to the consciousness of middle C. However, in real life this simple act is part of a more complex act in which I judge that the heard tone exists (as Brentano prefers to say, I affirm its existence) and, further, I feel a soaring delight in the affirmed heard tone. There may be a problem lurking here, as to whether the feeling attaches to the object or the act or both. The point to note, in any event, is that we need to articulate the structure of this consciousness into several components. And the act itself plays two roles: first, it is that which is directed externally toward middle C; second, it is itself that toward which it is directed internally. Thus, the act of consciousness includes a consciousness-of-itself "without any...further multiplication of entities," that is, without a second act directed toward the first.

Strictly speaking, on Brentano's analysis, the *object* of consciousness is a complex, a "double object," comprising as *primary object* (here) the tone and as *secondary object* the act of consciousness itself. The act of consciousness consists in its directedness toward this complex object. Because the act intentionally "contains" its object, the act intentionally contains itself – in a specific way. That way is the way of inner consciousness.

Brentano's analysis is awkward because he treats directedness as "containing"-an-object. Suppose instead we treat it, as I prefer, as an intentional relation with different modes and modalities of presentation defining "content."[15] Then the component characters of the act of hearing are distributed over the intentional relation in certain ways, and not over the object of consciousness. We might then revise Brentano's analysis into a form closer to my own analysis as follows:

Consciously I
[1] (am <u>aurally presented</u> this middle C tone and
   <u>judge</u> that it exists and <u>feel</u> that its Steinway resonance is delightful) and
[2] (am <u>herein presented</u> this very act of consciousness and
   <u>judge</u> that it exists and <u>feel</u> that it is pleasurable).

Inner consciousness, in this neo-Brentanian analysis, consists in the act's reflexive directedness through the indexical content "this very act of consciousness" in the position of internal presentation marked as "<u>herein presented</u>." The full character [1] of the consciousness of the tone is indicated by this internal presentation of the act itself. But the full character [2] of inner consciousness includes, for Brentano, not only that presentation of the act but also the judgment and feeling about the act – because inner awareness takes the act itself to exist and (in the case described) to be pleasurable. And the full character of the act, comprising [1] and [2], is indicated by "this very act of consciousness" within [2]. (The problem remains of whether there is pleasure both in the consciousness-of-the-tone [1] and in the consciousness-of-the-hearing [2]. But that problem is not our present concern.)

What should we learn from Brentano's study of inner consciousness? The key lesson, I think, is this: by parsing consciousness into several features of intentional structure, Brentano has shown that inner consciousness may be built into the structure of consciousness itself, as he says, "without any further complication or multiplication of entities." He has given us a model of how this might work, even if we reject the details of his analysis (as we shall see). With that model in mind, let us turn to more recent efforts to analyze consciousness.

Higher-Order Monitoring

Today's philosophy of mind – in the tradition of analytic philosophy since the 1950s and of cognitive science since the 1970s – is rediscovering the issues of consciousness that so engaged Brentano, James, and Husserl. This latest phase of consciousness theory is surveyed (*sans* phenomenology) in the anthology *The Nature of Consciousness* (1997), edited by Ned Block, Owen Flanagan, and Güven Güzeldere. Our purpose here is well served by the contrast of views gathered there in the section on "Higher-Order Monitoring Conceptions of Consciousness" and overviewed by Güzeldere in the closing essay (pp. 789–806).

In these recent discussions, inner awareness (imposing the term I prefer) is analyzed as "higher-order monitoring": the mind monitors its own activity in a distinct higher-order state – where all this is realized in the brain. This second state is higher-order because it is about the mental state it monitors, the latter being of first order, that is, about something other than a mental state. Two types of higher-order state have been posited as doing the monitoring: higher-order perception and higher-order thought. Higher-order perception is like perception but directed toward a given mental state rather than, say, a frog before one's eyes. Higher-order thought differs in that it is a propositional attitude of thinking about the given mental state rather than a kind of perception of the state.

David Armstrong offered this characterization of higher-order *perception*:

What is it that the long-distance truck driver lacks [when driving by skill without "consciousness" of what he is doing]? I think it is an additional form of perception, or, a little more cautiously, it is something that resembles perception. But unlike *sense*-perception, it is not directed toward our current environment and/or our current bodily state [proprioception]. It is perception of the mental. Such "inner" perception is traditionally called introspection, or introspective awareness. (Block et al., 1997: 724)

In the 1950s Armstrong, U. T. Place, and J. J. C. Smart launched today's materialist philosophy of mind with the identity theory, holding that a mental state is identical with a brain state (token states, not state types). It is natural on this model to suggest that the brain not only processes information about the environment, say, in seeing the road ahead, but also processes information about its own states, say, the brain's processing the sensory information about the road. This higher-order brain process is, then, what Armstrong proposes as defining consciousness, as when the truck driver becomes aware of his driving: his brain now "perceives" its state of seeing the road ahead. As the visual nervous system tracks interactions of the eyes with light, and the tactile system tracks interactions of the skin with physical pressure, so another part of the nervous system tracks neural interactions within the brain, which we experience as introspective awareness.

Since the 1970s, with the rise of the computer model of mind, the identity theory has given way to functionalism, holding that a mental state type is a functional state of a brain – or, indeed, of a computational system realized in a physical system different from a brain. Inner awareness would then be a type of higher-order information processing, a function of monitoring another mind/brain/computer activity (which has its own function). (See William G. Lycan in Block et al. 1997: 755–71.) On this paradigm, what makes a physical state *mental* is the type of causal interaction it mediates, as between a human nervous system and its environment; and what makes a mental state *conscious* is this type of higher-order monitoring of its own activity, a type of causal interaction within its own neural states.

But our concern is phenomenological: what is the *form* of inner awareness on the higher-order perception model? (We are seeking not a logical form of words, but a phenomenological form of mental states.) On the inner-perception model, introspective awareness is a second-order intentional state, which is to say it is directed toward a first-order mental state. And it is in some way like perception but it is not a sensory experience. In an idiom closer to my own, we might describe such "inner perception" as follows:

I now see that stretch of road ahead and (simultaneously)
I now <u>introspectively observe</u> that I am now having this very experience.

The content "this very experience" directs the introspective awareness toward the visual experience it is monitoring, which lies in the immediate mental context of the introspection. In this respect the awareness is like perception; it is a form of observation. But the character of the introspective awareness is not seeing or hearing; the sensory character belongs rather to the visual experience. What the truck driver lacks when driving along "unconsciously," on this model, is this further observation of his seeing the road ahead.

An alternative model distinguishes higher-order *thought* from higher-order perception, while remaining within the materialist theory of mind (and presumably specifying the causal functionality of a mental state). In this vein David M. Rosenthal argues, in discerning detail, for the view that inner awareness (again imposing my term) is a form of *thought* rather than perception (Block et al. 1997: 740ff.). Perception is too restrictive, even inappropriate, for introspection, Rosenthal argues, because its objects are restricted to what is sensible and has sensory qualities, whereas introspection ranges over all kinds of mental activity, with no such restrictions. Thus, he writes:

> We are conscious of something, on this [higher-order thought] model, when we have a thought about it. So a mental state will be conscious if it is accompanied by a thought about that state. The occurrence of such higher-order thought (HOT) makes us conscious of the mental state; so the state we are conscious of is a conscious state. Similarly, when no such HOT occurs, we are unaware of being in the mental state in question, and the state is then not a conscious state. The core of the theory, then, is that a mental state is a conscious state when, and only when, it is accompanied by a suitable HOT. (Block et al. 1997: 741)

But what is the *form* of the higher-order "thought" that makes a mental state conscious? It is not a sensory perception, describable in my idiom by something like "I see this very experience" or "I see that this experience occurs" (we do not *sense*-perceive mental states). It is instead a *propositional* mental attitude. But what type of attitude is it (judgment, imagination, etc.) and what form of proposition is its content? Rosenthal proposes:

> The requisite content is that one is, oneself, in a particular mental state. If one doubts or wonders whether one is in a particular mental state, or desires, hopes, or suspects that one is, that plainly will not make the state a conscious state.
> ...Nor will one's mental states be conscious if accompanied merely by a dispositional higher-order mental state.... To be transitively [= intentionally] conscious of something, therefore, we must have a thought about it in a relatively narrow sense: It must be an assertoric, occurrent propositional state. (p. 742)

Rosenthal's account may be summarized, then, as follows (condensed from pp. 741–43):

> One's mental state is conscious (intransitively), and thus may be conscious (transitively) of something, if and only if it is accompanied by an occurrent, assertoric propositional attitude whose content is that one is, oneself, in that particular state and which is thereby conscious (transitively) of that state but not itself conscious (intransitively).

Let us articulate the higher-order thought model in terms closer to my own. An assertoric, occurrent propositional state is a *judgment* – as opposed to a belief, which is dispositional, or a doubt, which is not assertoric, or a perception, which is sensory. And the relevant content is indexical, referring to "oneself" and to a particular mental state. In my idiom, the form of higher-order thought that makes a mental state conscious, on Rosenthal's analysis, may be described as follows:

I now judge that I am now in this mental state.

Because this thought accompanies the mental state as it occurs, my overall mental state may be described, in a simple case, as follows:

I now see this frog and (simultaneously) I now judge that I am now in this mental state.

The underscored phrasing ascribes the form of higher-order thought that makes the visual state itself conscious. The adverb "now" emphasizes that the visual state and accompanying judgment about it are occurring at the same, the present time. A mental state that is not conscious, then, is not accompanied by such a higher-order thought. Its description would be, say, "I now see this frog."

Compare this analysis with a Brentanian analysis recast, in this case, as follows:

Consciously I now see this frog and judge that I am now in this very mental state.

The chief difference is that for Brentano there is only one mental state and its structure includes a form of inner awareness, ascribed by the underscored phrasing, that is an integral part of that one mental state. The form "in this very mental state" lies within the scope of both "judge" and "consciously" (projecting the phenomenological form of the awareness from the logical form of the description). By contrast, for Rosenthal there are two mental states, where the second accompanies the first but is not an integral part of the first. The "and" joins two mental acts on this model, whereas on the Brentanian model the "and" joins two forms of directedness within one act – the one act is directed primarily toward its object and secondarily toward itself.

Brentano himself, pressing for a more complete psychology, noted the feeling that accompanies any mental state (on his hypothesis). Indeed, this element must be addressed in a fully adequate analysis, but let us simplify to focus on the form of the inner awareness and how it relates to the mental state it makes conscious. Then we note that Rosenthal joins Brentano in holding that judgment is the type of attitude appropriate to inner awareness. And, given recent logical theory, we can add the indexical form of proposition as the content (there is more here than meets the eye).[16] Like Hume, Brentano rejected the self and so would omit a form like "I am in . . ." But there is a self (whatever the ontology, I exist), and the indexical content "I" represents it. So let us allow, for present purposes, the content "I am in . . ." or "I now see this frog."

What shall we make of these higher-order monitoring models of inner awareness? It is indeed natural in the age of neuroscience and computer science to say that the brain monitors itself in a kind of "perception" of its own activity. And in the wake of modern logic it is natural to look for both a form of judgment about the conscious act and an indexical form of content that represents the act. Indeed, the indexical content "this very mental state" invokes the respect in which higher-order "thought" is already "higher-order perception" for Armstrong's purposes (perception is a form of indexical awareness).[17] So if we follow out the two lines of analysis drawn from Armstrong and Rosenthal, we find the basic analysis that we compared with an updated Brentanian account. But which of these two broad views is preferable, the Brentanian model of inner consciousness or the model of higher-order monitoring?

Inner Awareness without Higher-Order Monitoring

On the modal analysis I proposed originally (Smith 1986, 1989), inner awareness is internal to a conscious state, as Brentano prescribed, but does not involve higher-order monitoring (in judgment, perception, or recollection); it takes a different form. The more recent discussions (in Block et al. 1997) show what is at stake.

The problems facing the higher-order monitoring theory of consciousness were well detailed by Brentano and are addressed by Rosenthal in his careful discussion. First, there is the threat of regress. The response is to declare that only one level of higher-order judgment is needed: the articulated form of judgment is what makes the mental state conscious, and there is no further judgment about that judgment. I believe this is an appropriate response within the higher-order monitoring model. Second, there is the phenomenological observation that we do not *experience* a higher-order judgment (or inner perception) as distinct from the state it is supposed to make conscious. There I think lies the real rub. Rosenthal accepts this observation and even suggests (Block et al. 1997:

742) that the theory predicts it. However, we must be more sensitive to how we *experience* inner awareness.

Normally, we do not experience a *second* mental state of judging or inwardly perceiving about our ongoing conscious state. Inner awareness does not *accompany* a mental act as a second and second-order mental act. Rather, inner awareness is experienced as an *integral part* of a conscious experience: that is how we *experience* inner awareness. Brentano's analysis seeks to account for this fact of experience, while the theory of higher-order monitoring, whether thought or perception, does not capture this phenomenological fact.

Rosenthal addresses this worry when he asks (p. 744), "Must the relationship between mental state and HOT [higher-order thought] be closer than mere accompaniment? In particular, must we postulate a causal connection between the two?" A causal link, he concludes rightly, is too strong. However, the link we need is not causal; it is *formal*, a feature of phenomenological form. What is still missing in the higher-order thought model considered here is a formal structure that places inner awareness inside the mental state, rather than alongside it by accompaniment ("I see this frog <u>and</u> I judge that I see this frog").

But the problem with higher-order monitoring is not only that it lies in a second mental act we do not seem to experience; it is, if present, strictly *unconscious*. In my essay (1986) I posed the rhetorical question: how can an unconscious higher-order state make conscious a state it is about? Addressing my question explicitly, Rosenthal answers:

A state of consciousness can be a conscious state, that is, a state one is conscious of being in. Or it can be a mental state in virtue of which one is conscious of something [= intentional]. Adapting our terminology, we can call these intransitive and transitive states of consciousness, respectively. For a mental state to be conscious is for it to be a state one is conscious of being in. So a HOT [higher-order thought] can be a source of consciousness for the mental state it is about because the HOT is a transitive state of consciousness; it does not also need to be an intransitive state of consciousness. (Block et al. 1997: 743)

Rosenthal is right to distinguish a state's being conscious from its being conscious-of-something, or intentional. And, in the model at hand, he is right to say it is the HOT's being about the state that makes the state conscious. And he is right in noting (pp. 743–44), for example, that one may not be aware of hearing a conversation nearby until one's name is heard. But still my question remains, if I may recast it a bit differently: how can an unconscious intentionality directed toward a second intentionality make the second one conscious? The HOT is not a *consciousness*, or conscious "intention," of the given state, although it is an intention of the state. But if it is the HOT that makes that mental state conscious, then it ought to be somehow integrated into that consciousness. It ought to be a proper part of the given conscious state and so itself somehow conscious, something experienced in that mental state. Thus, we still have not captured something Brentano was onto in his own account of "inner consciousness."

Fred Dretske approaches the problem at hand from a different direction. While working very much in the tradition of cognitive science, Dretske rejects inner awareness as posited by higher-order monitoring theories of consciousness:

> What makes an internal state or process conscious is the role it plays in making one (intransitively) conscious – normally, the role it plays in making one (transitively) conscious of some thing or fact. An experience of *x* is conscious, not because one is aware of the experience, or aware that one is having it, but because, being a certain sort of representation, it makes one aware of the properties (of *x*) and objects (*x* itself) of which it is a (sensory) representation. My visual experience of a barn is conscious, not because I am introspectively aware of it (or introspectively aware that I am having it), but because it (*when brought about in the right way*) makes me aware of the barn. It enables me to perceive the barn. (Block et al. 1997: 785, emphasis added)

We return later to the possibility that consciousness does not always require inner awareness of the experience. For now, however, I want to extract from Dretske's discussion a different claim, a quasi-Brentanian claim about inner awareness. (Again, a state is "intransitively" conscious if it is simply conscious, and "transitively" conscious if it is conscious of something, or intentional.)

Dretske's theory of mind, here in the background, draws on information theory, holding that a mental state of an organism is a state relating it to its causal history and specifically to its transmission of physical "information."[18] In this vein, what makes a state mental is the form of information-transmission it realizes, and then what makes a mental state conscious is "the role it plays in making one (intransitively) conscious," namely, a role involving its causal reception of information from its environment. Overlaying this information-theoretic conception of mind, however, is a claim about the phenomenology of consciousness. Dretske holds (with Rosenthal) that consciousness comes in two *forms*: transitive and intransitive consciousness. I may be *conscious* – and this may involve my mental state's role in relation to my environment – or I may be *conscious of something*, say, in perception or thought or anger. The former is intransitive, the latter transitive. If you will: to be *conscious* is to be in a state of mind, often with a certain alertness, in which I am ready (say) to perceive consciously this frog or to judge consciously that this frog is about to jump.

On Dretske's non-Brentanian view, then, I may be *conscious* without performing any higher-order thought or perception about my mental state, and I may be *conscious of something*, as in consciously seeing this frog or consciously thinking it will jump, without any higher-order introspective awareness of my mental state. What makes my mental state conscious, on this view, is something else. Dretske recurs to its *role* (vis-à-vis environment) in making the state, well, conscious (and so ready for appropriate causal interactions with its environment). But we must distinguish causal role from phenomenological form. Then Dretske's contention boils down to the claim that what makes a mental state *conscious* is a feature more basic than that of inner awareness *of* the state. Whatever the causal role of consciousness, and given that role, its form *qua* conscious is different from that of transitivity, or intentionality: different, in particular, from any awareness *of* the conscious state. At least some conscious states, arguably, are conscious by virtue of this more basic feature. Fair enough. Still, we should make room in our phenomenological inventory for forms of consciousness that do involve inner awareness, even if other (perhaps lower-level) forms of consciousness do not. (We return to this point later.)

Our task, remember, is to analyze the *form* of inner awareness, the *phenomenological form* of the awareness that, by hypothesis, makes a mental state conscious, a form we *experience* in living through a conscious mental state.

Returning to our updated Brentanian model, we can now pose a question about the phenomenological form ascribed in Brentano's analysis – the form (in our idiom):

> Consciously I now see this frog and judge that I am now in this very mental state.

Why isn't the basic form of consciousness simply the form "consciously"? Why is an awareness *of* the state in a *judgment* that "I am in this very state" required to make the state conscious? Whether the second-order judgment merely accompanies the state (as in Rosenthal's analysis) or is bound into it structurally (as in Brentano's analysis), why is an act of *judgment* required to make the state conscious?[19]

Do we experience simply a character of consciousness, or do we experience consciousness embracing inner awareness within it? Brentano bound inner consciousness into consciousness of an object, as indicated in the neo-Brentanian analysis earlier. Brentano's intuition seems right, but the form of judgment seems wrong, even when bound into the complex form given previously.

Jean-Paul Sartre described consciousness as "translucent": I see *through* my consciousness, which is a consciousness *of* its object. Consciousness is at once a *conscious* experience *of* some object: it is "for itself," a prereflective "consciousness (of) itself," while being a consciousness *of* its object. This "consciousness (of) consciousness of something," Sartre says, is *constitutive* of consciousness, but it is not an additional consciousness. (See Sartre's doctrine of the prereflective cogito in the introduction to *Being and Nothingness*, 1943.) Sartre's description is suggestive and echoes Brentano's doctrine without adding an element of judgment. Still, his device of parentheses around "of" marks a distinction but does not adequately articulate the form of this "consciousness (of) consciousness."

As an alternative metaphor, instead of saying consciousness is "seeing" its object "through" itself, we might say consciousness is "hearing" its object "in" itself: *in a medium* that includes act and object in one movement. Hearing places me within the object of consciousness and within the experience, as the sound surrounds me, envelops me, flows through me – I am not looking at it from a distance. Nodding toward physics, we might say consciousness occurs in a "field" that ties act and object together so that the act feels itself in feeling its object.

Such metaphors seem to me on the right track, and they have their place in phenomenological description, rather like impressionist painting. What we need, accordingly, is an articulate model that places inner awareness *within* the consciousness of the object, without making it a separately focused higher-order cognition of that consciousness.

Inner Awareness as a Modal Character

We can now better interpret the *modal* analysis of inner awareness (which may seem rather cryptic in my 1986 and 1989 accounts). On that model, the form of a conscious experience is (staying with the old example):

<u>Consciously</u> I now see this frog.

It is important to factor the phenomenological structure of an experience so that "consciously" is articulated as a character distinct in form from the intentional character of presenting (say) "this frog." In order to mark out the formal role of this character "consciously," we need to develop a phenomenological distinction drawn sharply by Husserl.

Husserl distinguished two basic components in the "content" or phenomenological structure of act of consciousness.[20] In our simple case of seeing a frog, the "content" of the experience divides into the "thetic" or "positing" character of seeing (as opposed to judging, wishing, etc.) and the "sense" (*Sinn*) of the object, embodied in its presentation as "this frog." Similarly, Searle distinguishes the "psychological mode" from the "propositional content,"[21] say, in seeing that this frog is jumping. In the terminology I settled on (1989), the *mode* of presentation in a mental act carries the sense of the object ("this frog" or whatever), whereas the *modality* of presentation in the act carries the mental act type ("see" or "judge" or whatever). However, there are further qualifications in an act's thetic character or modality. I can see clearly or dimly; I can listen attentively or inattentively; I can desire passionately or casually; I can think quickly or slowly; and so on. On the modal analysis of consciousness, accordingly, a specific thetic or modal character "consciously" further qualifies the character of vision (or judgment or desire, etc.). The point I am pressing right now is that, within the formal structure of a mental act, the character "consciously" should be placed in the modality of presentation and not in the mode of presentation. The character of consciousness per se is then distinguished from the object-directedness of a consciousness. (By the way, observing this distinction in phenomenological structure helps us avoid any tendency to collapse the character "consciously" into another form of "consciousness-of-x"[22] in some higher-order cognition: the "of"-ness in the act lies elsewhere.)

In my modal model of inner awareness (as recounted earlier), the form "consciously" was factored into two components indicated in the phenomenological description:

<u>Phenomenally</u> <u>in this very experience</u> I now see this frog.

Here we find a way of articulating the form of consciousness that embraces inner awareness without pressing it into the mold of a separate judgment or inner perception, the mold of higher-order monitoring. Let me explain.

The form "phenomenally" covers what philosophers call the "raw feel" of an experience, its subjective quality, "what it is like" to have the experience, the way it is experienced or lived through. Here lie the sensory qualia in a perceptual experience. However, emotions also have their own peculiar qualia or subjective character: note the difference between feeling joyful and feeling sad. But qualia do not end with sensation and emotion. Conscious thinking "feels" a certain way. And so does conscious intention or volition. So every conscious mental act or state, we should recognize, has its *phenomenal* character. (Of course, this use of "feels" is not tied to emotional "feeling.")

Now, there is nothing in the phenomenal character of an experience that articulates an inner awareness of the act as such. My proposal is that there is a further *reflexive* character, the character "in this very experience" – or "herein" – that articulates awareness *of* the experience itself. But there is nothing in this reflexivity that articulates phenomenality; a mental act could intimate itself reflexively without yet being phenomenal. These two characters are interwoven in consciousness, in inner awareness, on the model at hand. Here lies inner awareness, in the form "phenomenally in this very experience."

The character "phenomenally" modifies the form of the whole mental act. And within the scope of that character, the form "in this very experience" – or "herein" – reflexively indicates or intimates the act itself. But nowhere in this structure is there the form of separately observing or judging that the act is occurring, or that I am in that mental state. Thus, within the overall form of the conscious experience there is a form of inner awareness of the act, in this reflexive character. Shall we call this a "secondary intentionality," with Brentano? That would be misleading, because it is not a part of the act's directedness toward its object; nor is the object of consciousness itself a complex including (here) the frog and the act. "In this very experience" is not a separate mental act, nor a component act. Rather, it is a character of the act, of the way the act is executed.[23]

The reflexive character "in this very experience" is modified by (falls within the scope of) the phenomenal character "phenomenally." In that way the inner awareness of the act is itself conscious and experienced or "felt" in living through the act. So inner awareness is neither unconscious nor a further act accompanying the given act.

On this modal analysis, then, the form "this frog" is part of the *mode* of presentation of the act's object, prescribing the object intended, while the form "see" is part of the *modality* of presentation or intentionality in the act. The character "consciously" is a further part of the *modality* of presentation or intentionality. And so the inner awareness articulated by the form "in this very experience" is part of the modality of intentionality: part of what makes the act conscious, within the character "consciously." Yet "this very experience" does not occur in any form of judgment holding that I am having "this very experience," whether a separate judgment (Rosenthal) or an internal judgment (Brentano). Nor does it occur in any other form of higher-order monitoring.

In short, we have here an analysis of phenomenological form that resonates with Brentano's view of inner consciousness but avoids the problematic features of higher-order monitoring, including Brentano's own version of an internal higher-order judgment.

Temporal Inner Awareness

In lectures published as *On the Phenomenology of the Consciousness of Internal Time* [inneren Zeitbewusstsein] (1991/1893–1917), [24] Husserl developed a quasi-mathematical analysis of our consciousness of the "inner" passage of time. According to Husserl, our "inner consciousness" of our passing experience – what I am calling inner awareness – automatically falls out of the structure of our consciousness of time: our temporally flowing consciousness of temporally flowing events both external and internal to our consciousness. Husserl's analysis of time consciousness seems to imply, then, that inner awareness in our sense can be analyzed more closely in terms of the temporal structure of consciousness. Let us explore this temporal model of inner awareness in light of our preceding considerations.

So long as we are conscious, we experience a temporal flow of sensation, a continuous flow of sensory awarenesses in visual, auditory, tactile, and kinesthetic perception, along with taste and smell. Consider a simple experience of hearing a melody, say, "Yankee Doodle." Each phase of this conscious process is complex. For at one moment I hear one present tone as part of that familiar melody, both retaining the sound of certain past tones and anticipating the sound of certain future tones to come. On Husserl's analysis, the structure of my hearing this tone in "Yankee Doodle" is a fusion of: (1) *hearing* this present tone (E above middle C), (2) *retaining* some past tones (C, C, D), and (3) *anticipating* some future tones (C, E, D). The whole experience includes as concurrent parts my current hearing of E, the sequence of my "retentions" of past tones just heard, and the sequence of my "protentions" (Husserl's neologism) of future tones about to be heard, if we assume the melody continues. And there are mutual dependencies among these phenomena: my hearing E in "Yankee Doodle" depends on those retentions and protentions, and vice versa. (Husserl's ontology of dependent parts would come into play here.) Retention is different from recollection or memory; retention is very short term and an integral part of perception, not a distinct act of recollection. And protention is similarly an integral part of perception, as distinct from expectation proper.

As former assistant to the mathematician Weierstrass and so attuned to the infinitessimals of the calculus, Husserl treated the flow of auditory consciousness as a continuum of point-sounds in a temporal curve presented in a corresponding continuum of point-hearings. A more realistic account of hearing a melody divides the experience into discrete tone hearings, each with its presented duration, volume, and thus rhythm. Needless to say, a conductor's hearing his symphony's performance of a Beethoven work is a very complex perception, and a technician mixing the sound of a synthesizer in a recording studio is also hearing a highly structured sound, one that appears with a mix of continuous phrasings. Husserl's analysis is thus greatly idealized, yet it remains instructive.

Now, according to Husserl, retention has a "double intentionality": each retention of a past tone includes both a retention of the *sound* just heard and a retention of the just-past *hearing* of the sound – that is, a primary and a secondary intentionality (thus applying Brentano's doctrine to retention). And each protention similarly includes an anticipation both of the sound and of the hearing of the sound. (Husserl distinguished more clearly than Brentano the hearing, the sound, and indeed the appearance of the sound.) So my experience in hearing E, the fourth tone in "Yankee Doodle," is a complex whole consisting of hearing the present tone while interdependently retaining past tones and past hearings and anticipating future tones and future hearings.

From this complex phenomenological structure, we can carve out a distinctively Husserlian analysis of inner awareness as *temporal awareness*, a successor to Brentano's effort at appraising "inner consciousness." The analysis I see runs as follows.

At the moment I hear E, I have no concurrent secondary intention of my hearing E. My experience includes at that moment, according to Husserl, a "primary impression" of E but no secondary intention of that auditory presentation or hearing of E. Part of my experience of hearing E, however, is the pattern of secondary retentions of past hearings and secondary anticipations of future hearings. On the Husserlian analysis I am suggesting: my *inner awareness* of my experience of hearing that tone is *constructed* from this pattern of *secondary* retentions and protentions. Interestingly, there is a "blind spot" (or "deaf spot") in this inner awareness, as my experience includes no current awareness of hearing E. Nonetheless, because this experience is itself part of an ongoing process of auditory consciousness, in the next moment there will occur a secondary retention of the just-past hearing of E. And so it goes, the flow of auditory consciousness, including as a dependent part a flow of secondary awarenesses of the process of consciousness. According to this Husserlian analysis, my inner awareness of my ongoing experience of hearing "Yankee Doodle" played on this piano reduces to this ongoing flow of secondary awarenesses.

Husserl struggled, in the time consciousness lectures (1991/1893–1917), to articulate this inner consciousness of temporally flowing experience in a way that would avoid suggesting any form of higher-order consciousness:

There is one, unique flow of consciousness in which both the unity of the tone in immanent time and the unity of the flow of consciousness itself become constituted at once. As shocking (when not initially even absurd) as it may seem to say that the flow of consciousness constitutes its own unity, it is nonetheless the case that it does.... Every "retention" possesses a double intentionality: one serves for the constitution of the immanent object, of the tone.... The other intentionality is constitutive of the unity of this primary memory [of the just sensed tone] in the flow. (§39, 84)

And still more precisely: In its process of being continuously adumbrated in the flow, it [the retention] is *continuous reproduction of* the continuously preceding phases. (p. 390, from a text amplifying the text of a preceding quotation)

Husserl finds his results here "scandalous" (in the words of his translator John Brough, p. XIX), asking how "constitution" is possible (p. 390). The scandal being that, on this analysis, temporal consciousness "constitutes" both its object-in-time and itself-in-time.

In Husserl's neo-Kantian idiom, consciousness is said to "constitute" its object: not that it creates the object, but that it "intends" or presents the object through a certain meaning.[25] Here Husserl claims that consciousness also simultaneously "constitutes" itself, that is, its own unity in time. And yet, later in the text, Husserl insists that this self-"constitution" does not form a further act:

Every act is consciousness of something, but there is also consciousness of every act. Every experience is "sensed," is immanently "perceived" (inner consciousness), although naturally not posited, meant (to perceive here does not mean to grasp something and to be turned towards it in an act of meaning)....

This present, now-existing, enduring experience ... is already a "unity of inner consciousness," of the consciousness of time; and this is precisely a perceptual consciousness. "Perceiving" here is nothing other than the time-constituting consciousness with its phases of flowing retentions and protentions. Behind this perceiving there does not stand another perceiving, as if this flow itself were again a unity in a flow. (Appendix XII, 130; substituting "inner" for "internal" in the translation)

Husserl is explicitly responding here to the familiar threat of infinite regress. Note that he insists on the term "perception" for "inner consciousness," yet explicitly denies that it is what philosophers today call a higher-order perception. The "scandal" is really the problem of locating inner awareness within the unified experience while keeping it from looking like a higher-order "perception" of the experience that steps back and separately "intends" the experience.

We need a more effective way of articulating the *form* of "inner consciousness" that Husserl is trying to describe. Along these Husserlian lines, may I suggest, a phenomenological description of the whole experience might look like this:

> Phenomenally (I now hear this present tone E in "Yankee Doodle" and [interdependently] retain (i) past tones ... and (ii) my past hearings thereof and [interdependently] anticipate (i) future tones ... and (ii) my future hearings thereof).

The character "phenomenally" distributes over the whole structure of the experience, and then, in Brentanian fashion, the pattern of secondary retentions and anticipations makes up my *inner awareness* of the whole experience. Thus, in Husserl's hands inner awareness resolves or decomposes into this pattern of retentions and protentions. Inner awareness itself is thus reduced to *temporal awareness* of flowing consciousness.

This analysis is elegant. But is this temporal structure the proper *form* of inner awareness as such? I think not, although inner awareness may rightly be said to supervene on temporal awareness.

Inner Awareness Supervening on Temporal Awareness

When I see a frog, my experience includes a visual presentation of a frog-like pattern of colors and shapes and textures. But the salient content in my experience – the salient structure in the mode of presentation of the object – is "this frog," or "this red tree frog." That content, we may say, supervenes on the visual registration of details of color, shape, texture – of the frog as presented. But the content "this red tree frog" does not decompose semantically into a content specifying the pattern of colors, shapes, and so on: that is the lesson of the failure of sense-datum theories of perception, and of careful phenomenological analysis. Similarly: when I hear the melody "Yankee Doodle," my inner awareness of my perception carries the content – part of the modality of presentation – "in this very experience." My inner awareness, bearing that content, supervenes on my temporal awareness consisting in a pattern of retentions and protentions, but the content "in this very experience" does not decompose semantically into a content specifying the pattern of retentions and protentions (like temporal sense data). Accordingly, the contents that define respectively *inner* awareness and *temporal* awareness are distinct contents with different phenomenological tasks.

The phenomenology of my experience of hearing "Yankee Doodle" should then articulate separately both inner awareness and temporal awareness. Along the lines we have been pursuing, I propose the following (rather cumbersome) phenomenological description:

Phenomenally in this very experience
(I now hear this present tone E in "Yankee Doodle"
and [interdependently] retain (i) past tones...and (ii) my past hearings thereof and [interdependently] anticipate (i) future tones...and (ii) my future hearings thereof).

My inner awareness of my experience, in virtue of the modal character "in this very experience," spreads holistically over my temporally flowing experience of hearing "Yankee Doodle" including the present tone E. But the reflexive content "this very experience" indicates my experience directly, without articulating the complexity of temporal awareness that Husserl appraised. Accordingly, the form of inner awareness, "in this very experience," is distinguished from the form of temporal awareness in the pattern of retentions and protentions. Yet the character "in this very experience" itself *depends on* the temporal awareness comprising retentions and protentions. For, without that integral temporal awareness of the ongoing experience, I could not (as Husserl says) "sense" the ongoing experience, albeit simply as "this very experience."

Whatever else we experience, we experience a temporally structured flow of sensory perception along with thinking, imagining, wishing, willing, and the like. This flow of sensory experience (including kinesthesis and bodily control) is a deep and basic current in our stream of consciousness, above which flow our various other intentional activities. And that basic sensory flow, let us grant, has a structure very much like that Husserl analyzed as temporal awareness. Supervening on that temporal awareness, however, is a distinct *formal* feature of consciousness, which is inner awareness proper. There is the form of "inner consciousness" that Brentano and Husserl sought to explicate.

That said, the whole experience – bearing the whole content articulated by our complex phenomenological description – is itself flowing temporally, and its aspect of inner awareness is an integral part of the temporally flowing experience. Thus, inner awareness is itself temporally manifest, as the pattern of secondary retentions and protentions points toward the temporal extent of the whole experience, of which inner awareness is an integral part. To recognize inner awareness as part of the *modality* of presentation, I would suggest, helps to dispel the sense of mystery about the "self-constitution" of consciousness through "inner time-consciousness."[26]

Consciousness as Modality

When I spoke of the *modality* of presentation in an intentional act (1986, 1989), I alluded to a conception of intentionality as a form of modality. In the 1960s Jaakko Hintikka analyzed the logic of sentences like "I believe that *p*" and "I see that *p*" as similar to the logic of "necessarily *p*" or "it is necessary that *p*." Where the traditional metaphysical notion of modality covers necessity and possibility, Hintikka's logic brought intentionality into a similar mold. The logical analogy frames the following truth conditions:

> "Necessarily *p*" is true in a world W if and only if "*p*" is true in all logically possible worlds logically alternative to W.

> "*a* believes/sees that *p*" is true in a world W if and only if "*p*" is true in all intentionally possible worlds intentionally alternative for *a* to W.

In Smith and McIntyre, *Husserl and Intentionality* (1982), this modal semantics was transformed into a modal model of intentionality per se (as opposed to the interpretation of sentences ascribing intentional propositional attitudes). On that modal model, an act of thought or perception is intentionally directed not simply toward an object in the actual world but toward a variety of objects in appropriate situations or worlds defined by the content (and "horizon") of the act. That is, the act's intentionality consists in this pattern of directedness reaching into a variety of conceptually or intentionally possible situations (while based in a home world W, normally the actual world).

The modal analysis of consciousness begins with the logical form of the phenomenological description "Consciously I see this frog" or, *pace* Hintikka,

"(Consciously I see that) (this is a frog)."

Here the modifier "consciously" is parsed as an adverb within the modal operator "I see that" and then factored further in

"((Phenomenally in this very experience) I see that) (this is a frog)."

So we use the logical form of the modal operator to articulate inner awareness, along the lines explained previously. The assumption is that this form of language articulates the form of experience we are after. Accordingly, we call the phenomenological form "consciously" a *modal* structure of intentionality.[27]

Looking at the possible-worlds explication of intentionality as modality, we can see a striking difference between the modal model of consciousness and higher-order monitoring models. On the latter, the act A is itself the *object* toward which a certain higher-order act H is directed: that is the relevant fact that makes A conscious. However, on the modal model, there is only one act A – with no accompanying act H – and A is directed only (in the case at hand) toward a situation involving a frog. Thus, the content "this is a frog" in A directs my experience A in W toward a situation S′ in a world W′ (intentionally possible and alternative for me in A to W), where S′ consists in a certain object's being a frog, namely, the object visually before me in W.[28] A's intentionality consists in this pattern of directedness toward frog situations in the alternative worlds. Now, the content "consciously" in A plays a completely different role. It does not resolve into the intentionality of a second act H (directed toward A in its alternative worlds). Nor does it prescribe the object toward which A is directed in a given alternative world (perhaps placing A alongside a frog situation in a double object of A, as Brentano proposed). Rather, the character "consciously" modifies, or modalizes, the presentation in A (whose pattern of directedness is toward frog situations in alternative worlds).

Assume that "consciously" resolves into "phenomenally in this very experience." The higher-order content "this very experience" in A in W reflexively indicates A in W. And A in W serves as an anchor for indexical presentations such as A's presentation of "this frog [actually now here visually before me *in this very experience*]."[29] But while the content "this very experience" may be implicit in other forms of indexical awareness, including presentation of "this" or "this is a frog," its role in A is not to place A as part of the *object* of intentionality in A. Rather, its explicit presence in the *modal* character of A helps to define the character of inner awareness in consciousness, according to the theory at hand. But this role calls for further explication than we can provide within the familiar possible-worlds framework. The extensional analysis of intentionality in terms of possible worlds is helpful in bringing out the pattern of directedness of a mental act but is not well suited to bringing out this phenomenological character within the modality of presentation.

To develop the modal model of consciousness further, and so to add warrant to the modal analysis of inner awareness, we would need to explore the ontology of modality along with that of intentionality. With that discussion we would move from the phenomenology of consciousness – our present concern – to its ontology.

Levels and Biases of Consciousness

It seems obvious that there are different levels of consciousness. Just consider human consciousness as it varies from hunters, farmers, and carpenters to mathematicians, musicians, and writers to Zen masters. How does inner awareness distribute across such variations? Does the form of inner awareness occur in all levels of consciousness?

If we place our own species in the variety of biological species on the planet (let us not speculate here on extraterrestrial consciousness), it is obvious that there are different levels of development of mental activity, and such gradations ought to be found in consciousness itself. Surely then the lowly snail has a low level of consciousness in its sentient response to things it encounters: the rock is to climb, the leaf is to eat. Behaviorist and functionalist approaches to animal behavior would discourage us from speculating on "what it is like" to be a snail. Rightly so, if the consciousness enthusiast were inclined to wonder what the snail is "thinking" (reflectively?). Moreover, it would be inappropriate to speak of the "first-person" perspective of a snail, because the snail is not a "person" (and it will always be the "other"?). Still, the snail is somehow conscious of its environment and its activity.

Ants and bees perform intricate tasks in a complex society of their kin. Their specific movements and reactions are surely not founded in reflection upon the challenges at hand. Biology, nature's genetic engineer, controls most and probably all of their activity, and their "culture" is entirely wired into their social activity in building and maintaining their colony. Yet some of their sentience and action surely involves, like the snail, a low level of consciousness.

And then there are plants, which are said to respond to classical music, without recognizing Mozart. Is their response to water, light, even sound imbued with a strain of consciousness? Who knows?

Quite different, in any event, are paradigm human experiences of consciously thinking, perceiving, desiring, intending. These are the cases philosophers began with, notably and rightly, in the evolution of rationalism (Descartes), empiricism (Locke), the Enlightenment (Voltaire, Jefferson), German idealism (Kant, Hegel), Austrian logically minded phenomenology (Bolzano, Brentano, Husserl), and French existentialism (Sartre, Merleau-Ponty, Beauvoir) – times of high reflection in our "modernist" era. Philosophy packed into these human forms of consciousness a high degree of self-consciousness. Psychoanalysis (Freud) articulated still more complex types and levels and dynamics of self-consciousness: distinguishing conscious, preconscious, unconscious, and repressed mental states of belief, emotion, and motivation.

And so, surely, in the variety of human activity, we experience forms of consciousness embracing inner awareness as analyzed previously: that has been the theme of this chapter.

Still, we must not press every consciousness into the mold of "inner consciousness," as important as it is in certain higher activities of consciousness. For we also experience very basic modes of consciousness that lack "self-consciousness." Heidegger's phenomenology featured our habitual everyday "comportment" (*Verhalten*, or "relating"), as in hammering a nail, and Heidegger resisted Husserl's emphasis on "consciousness" (*Bewusstsein*, or being known). Similarly, Armstrong stressed the mental state of the "unconscious" truck driver ably driving down the highway. Now, I think we should not abandon the word "consciousness" in these cases, even if it came to us originally as a semitechnical term of modern philosophy. Instead, we should articulate different *levels* of consciousness. I am not *unconscious* when I am unselfconsciously hammering a nail, or driving down the highway, or choosing to hit the tennis ball crosscourt rather than down the line, or even when I am moving mentally to the next step in writing a proof in a logic class, or intently phrasing the next point in this essay. Nonetheless, in these experiences I am not particularly conscious of my mental activity, or of myself. Inner awareness seems to be lacking in some of these states of mind, even if they are conscious. Indeed, we should allow that some of our perfectly human experiences are, as it were, low-level sensory and motor activities that are conscious but lack a proper form of inner awareness. To develop this point calls for close phenomeno-psycho-neuro-logical analysis, perhaps seeking corroboration of the phenomenology from studies of our brains and our evolutionary biology: but let us make room for the point here.

On the view now emerging, inner awareness is an integral part of higher levels of consciousness, realized in humans and perhaps other animals, but it is not present in lower levels of consciousness in humans and other animals. If so, we must distinguish forms of consciousness with inner awareness from those lacking it. This distinction runs counter to the traditional Cartesian-Lockean-Brentanian conception (which I have followed in 1986, 1989). Yet it may be naturally parsed from my initial analysis (redeployed earlier): lower levels of consciousness have the character "phenomenally" without the character "in this very experience," whereas these characters are both present (and fused) in inner awareness at higher levels of consciousness. The problem now is to explicate appropriate levels of consciousness and to specify where inner awareness occurs in various levels.[30]

Whatever the empirical facts of consciousness, in various forms of life, we may distinguish certain *formal* features of consciousness that define four basic levels of consciousness. Thus, we propose the following (partial) scheme of *categories* of consciousness in four levels:[31]

  1. Sentient-Motor Activity.
  2. Indexical Awareness.
  3. Conceptual Intentionality.
  4. Symbolic Intentionality.

The snail is engaged in *sentient-motor activity,* and so am I when I stub my toe on a rock and instinctively pull it back. Sensory awareness and purposeful bodily movement at this most basic level are conscious but lack inner awareness. No doubt my sentient-motor activity at this level is interlinked with far more complex mental activities than the snail's, yet both occur at this formal level of consciousness.

Indexical awareness begins with awareness of the distinctness between "this" and "that": between a state or entity "itself," "this here entity" ("self" without a sense of person), and an entity over against it, "that there entity." This second level of consciousness bears the protoform of intentionality, the distinction between act and object. Here we find the most basic form of inner awareness, carried in the consciousness of "that" object over against and distinct from "this" act.

Conceptual intentionality presents an object under a predicative concept, such as "frog" or "rock" or indeed "person." My simple visual consciousness of "this frog" involves elements of consciousness from the first three levels: my experience is a visual sentient activity, it is an awareness of "this/that [entity]" (as distinct from "this very experience"), and it is a conceptual intention or presentation of "[this] frog." Indeed, the conceptual presentation depends on the indexical awareness, which depends on the sentient activity. This level of consciousness assumes a conceptual apprehension of space-time and the place of experiences and objects in it.

Symbolic intentionality occurs when a consciousness involves symbolic concepts (some of them indexical) that have a grammar or syntax in a logical-linguistic symbol system, notably, a natural language like English or Chinese. When I consciously think "Brentano launched the phenomenology of inner awareness," my experience is laden with concepts that belong to and are expressed only within a proper language, namely, English – indeed, an ideolect of English philosophical discourse (following translation from German philosophical discourse). Again, when I consciously see "this Macintosh PowerBook G4 computer," my consciousness includes concepts that I could not have unless I were familiar with jargon of today's computer technology. The fourth-level symbolic consciousness in my experience depends on third-level conceptual presentations of shapes and colors and tools familiar in everyday human culture, and both depend on second-level indexical awarenesses of "this silver box . . . ," which in turn depends on first-level sentient visual consciousness.

Again, while we choose commonplace examples of these levels of consciousness, we are noting *formal* structures of consciousness, and the theory of these forms of consciousness is a formal theory that awaits empirical application. All that we need claim here is that a simple case of human visual consciousness, for example, can be factored along these lines into different elements. Where exactly does inner awareness occur in these levels of consciousness?

Where a mental activity is conscious, it has the character "consciously" or "phenomenally." This remains a primitive form of experience, and it covers the mental structures in the four levels of consciousness distinguished here. Inner awareness enters the form of consciousness most basically with indexical awareness of "this/that entity" as distinct from "this act." If we have a conception of "experiences," as we do in everyday life, then the form "in this very experience" is indexical and conceptual; this is the form we assumed in the analysis of our everyday awareness of our passing experience. And if one day we are able to discern our own neural activity, somewhat as an expert winemaker discerns the "cherry" and "chocolate" and "cassis" overtones in a Stags Leap appellation of California cabernet sauvignon, then inner awareness may include a symbolic-conceptual content: "Phenomenally in this very neural activity I see that poisonous red tree frog," where the form of inner awareness "in this very experience" is informed with the theoretical concepts of neuroscience.

The four-level scheme of categories of consciousness helps to explain some prominent criticisms of classical, broadly Cartesian philosophy of mind – and to explain where they go wrong. (i) As noted earlier, Fred Dretske has urged that consciousness is defined not by inner awareness but by a certain (causal) role. This view finds its natural (and naturalistic) place at low levels of consciousness, especially in sentient-motor activity. Indeed, at that level we want to address the causal interaction of a conscious organism with its environment.[32] Yet recognizing this level does not in itself cut inner awareness out of other forms of consciousness, especially at higher levels. (ii) Following Kant, Wilfrid Sellars has argued that perception cannot present "this" sensuous object without some sortal concept such as (in our example) "frog." Third-level perceptual consciousness would indeed take such a form, as in seeing "this frog." Yet it does not follow that there cannot be a lower level of perceptual consciousness of "this," where sortal concepts fail. When it is foggy, I may see "that" ahead (whatever it be), with no sortal concept informing my experience.[33] (iii) Drawing on Heidegger and Merleau-Ponty, Hubert Dreyfus has championed habitual human actions, describing states of what I would call "minimalist consciousness," as when I see a nail and hammer it in the familiar course of carpentering. I am consciously hammering the nail (I am not a zombie) but without any prominent consciousness – even minimal indexical awareness – of my action, myself, my hand, my hammer, my will, and without any conscious conceptual or symbolic judgment or volition about hammer, nail, cabinets, and so forth. Higher-level consciousness is lacking in this action, and inner awareness may be largely or even wholly lacking in such actions. Still, it does not follow that consciousness and inner awareness are lacking in other forms of human activity, at different levels of consciousness.

(iv) Ludwig Wittgenstein described "fundamental empirical propositions," contained in basic beliefs that are inaccessible to consciousness – not because they are repressed à la Freud, but because they are so basic to our human existence that we ordinarily have no need for them. Such beliefs would be symbolic or language-bound intentional states, but they do not involve consciousness, much less inner awareness (see in Chapter 5 the essay "Background Ideas," extrapolating from Wittgenstein's concerns in *On Certainty*). Such "beliefs" do indeed ground our epistemology. Yet it does not follow that consciousness with inner awareness does not inform other human thoughts, at the levels of conceptual and symbolic consciousness. (v) Donald Davidson has argued, on grounds of interpretation of others' beliefs and desires, that language is a prerequisite for intentional propositional attitudes. Only at the symbolic level of consciousness, however, do such intuitions and arguments have bite. In any event, it does not follow that lower levels of consciousness and intentionality require language.

Such observations, against the omnipresence of acute consciousness in human affairs, should not impel philosophy to abandon consciousness in an effort to get beyond Cartesianism and into high postmodernism, as a Richard Rorty might urge. Instead, such observations should lead us into careful discriminations of forms and levels of consciousness. In that spirit we proposed a four-level scheme of categories of forms of consciousness.

These four categories do not, however, complete the systematic categorization of forms of consciousness. They define certain forms of awareness, in four levels, but across these levels other categorial differences apply. A separate dimension of variation is that between three basic forms of mental act or state, which we might call forms of mental *direction* or *bias*.[34]

A. Perception   B. Thought   C. Volition/Action

In perception a being or organism receives information from its environment and so is affected by its environs. In volition a being or organism initiates action and so affects its environs. (Or, we may say, in action a being executes or carries out a volition.) In thought, in either judgment or imagination, a being represents or conceives things, but does not directly affect things and is not directly affected by the environment. In ordinary human affairs, thought, perception, and volition or action are interdependent. For what I see or do typically depends on how I think about things, and what I think typically derives from what I see and what I do.

In human perception, information is channeled in one of six sensory forms: seeing, hearing, touching, smelling, tasting, and kinesthetic sensing (of one's bodily movement). And these sensory modes of perception include qualitatively different forms of sensation: visual, auditory, and so forth. Sensation, we assume, is not a distinct form of consciousness or mental activity but is a feature (a dependent part) of an act of perception. And we assume this sensory quality in an act of perception occurs only where the act is conscious – unlike blindsight.

In human thought, intentional content follows the forms so much discussed by classical philosophy, featuring conceptual, propositional, logical forms familiar to us all and often shaped by our language. In judgment a state of affairs is not only represented but posited as existing or actual; in imagination, something is merely represented, with no stake in existence.

In human action, bodily movement is initiated by volition and guided by kinesthesis, typically framed by background ideas and expectations that define what one is trying to do. There is considerable intentional structure in action, beyond the scope of the present study. (See "Consciousness in Action" and "Background Ideas" in Chapters 4 and 5.)

Where does feeling or emotion enter the categorization of mind and consciousness? Brentano, we saw, held that a character of feeling was present in every mental act. If so, then emotion defines yet another dimension of variation in the form of mental states. There will be different types of feeling and different ways in which it enters the form of a mental act, especially a conscious one. One way emotion appears in consciousness is in connection with volition, as part of the motivation that leads into a state of volition, which initiates action. Where perception includes a quality of sensation, volition (and action) includes a quality of emotion. But emotion also informs thought and perception. We cannot take up these issues here, but we note their relevance to a category scheme for mind and consciousness.

If we put together the distinctions we have been observing, we may begin the categorization of forms of mind and consciousness with the following scheme:

A. Perception   B. Thought   C. Volition/Action

1. Sentient-Motor Activity
2. Indexical Awareness
3. Conceptual Intentionality
4. Symbolic Intentionality

Then different levels of consciousness characterize different biases of mental state, as the two dimensions of variation cut across each other. What I am suggesting in this brief outline is a categorization that organizes the kind of classifications observed in classical works of psychology by Brentano and James and extended in classical works of phenomenology by Husserl, Heidegger (terminology notwithstanding), Sartre, Merleau-Ponty, et alia. But the art of categorization should also reflect a wider purview, an ontology of mind and consciousness. This purview takes us much further afield. (A program of ontological categorization along such lines is outlined in Chapters 6–8.)

Forms of Consciousness in Review

Regrouping results, we can distinguish and organize a variety of forms of conscious (and unconscious) intentionality that we have encountered in the preceding discussion. Staying with simple perceptions, where consciousness is palpable, and focusing on forms of awareness, we distinguish the following forms of mental state or act, assuming their analysis in terms similar to those in the preceding discussion:

1. Unconscious vision. Blindsight is perception without consciousness, wholly unconscious vision.

2. Conscious vision without retention. The long-distance truck driver consciously sees the road ahead (and steers accordingly) but has no immediate memory of his perception (and steering). His perception includes consciousness but little or no retention of his ongoing vision (and steering) – and no awareness of seeing the road (and steering) while doing so. In some moments of his trip he may simply have no retention of what he saw (and did in steering) a few seconds ago, and yet he is and was consciously seeing the road.

3. Conscious vision without inner awareness. The well-focused carpenter consciously sees the hammer and nail (and hammers the nail) but has no inner awareness of his seeing (or hammering). His perception (and action) include consciousness but no inner awareness. However, the perception may include retention even without inner awareness. For, when asked what he was doing, he replies, "Hammering this nail," while of course seeing it, even though in doing so he has no awareness of seeing (and hammering), no concurrent holistic awareness of "this seeing [and hammering]."

4. Conscious vision with inner awareness. When I see this statuette beside my computer, I am quite aware of my "regarding" it as I think of its African origin. My consciousness in seeing it includes an

inner awareness of seeing it. And my perception includes retention, which plays a role in grounding the inner awareness as I look at it progressively from different sides, admiring the craftsmanship.

5. Conscious vision with inner awareness but without conceptual or symbolic content. When I see something ahead on the road in the fog, my perception is conscious, imbued with retention, and endowed with inner awareness. But no interpretation of what lies ahead is present in the content of my perception; there is no conceptual content and no symbolic content. I merely see "that," not "that stone" (conceptually presented) or "that Mercedes on the Autobahn" (symbolically or linguistically presented).

6. Conscious vision with inner awareness and with conceptual content but without symbolic content. When I see "this gizmo," part of the machinery inside my computer, my perception is conscious, imbued with retention and inner awareness, but lacking in symbolic content. I have formed a concept of this type of part, recognizing it. But lacking expertise, I have no properly symbolic words for it, words that would be part of the technical language used by electronic engineers in designing this part. ("Gizmo" is a dummy word, even if I use it while pointing at the part.)

7. Conscious vision with only background conceptual or symbolic content. When I see "this woman approaching," my perception is conscious, imbued with retention and inner awareness, and endowed with some conceptual content, but most of the relevant conceptualization (with or without symbolic linguistic articulation) is not present in my experience but resides in the background of my consciousness. I may recognize a human being, and his or her gender, and even the person herself, without any further background content (much less silent words) streaming through my experience.

8. Conscious vision with inner awareness and with symbolic content: When I see "this man approaching, Jacques Derrida, the father of deconstruction," my perception is conscious, imbued with retention and with inner awareness, and endowed with content informed by extant, grammatically well formed language, and specifically written language (as the object of perception would himself stress).

And, of course, there are more variations on the forms realized in states of consciousness. These cases, however, begin to lay out the variety of forms of consciousness, inner awareness, and retention that may be involved in perception.

In short, given the variety of structures we have found in different forms and levels of consciousness, we can begin to systematize the differences among the many cases that motivated different – and falsely competing – views about consciousness, its form, and its prevalence. We have, in other words, begun the systematics of consciousness.[35]

Notes

1. Descartes's view is expressed in the Sixth Replies: in the translation by Cottingham, Stoothoff, and Murdoch 1988: vol. 2. Descartes's Latin *"cognitione interna"* is translated by Cottingham et al. as "inner awareness" – a leap of interpretation. Remember that Descartes uses *cognitio*, or "thought," as a general term for all mental acts. When I wrote "The Structure of (Self-)Consciousness" (1986), I chose the term of art "inner awareness," which I later found in the Cottingham et al. translations and addressed in "The Cogito *circa* AD 2000" (1993; Chapter 1 in this volume). Although the translation is somewhat anachronistic, Descartes's argument is thoroughly contemporary with today's debate about higher-order thought making a mental state conscious – a view discussed here. My thanks to Alan Nelson for assistance on this point of translation and interpretation.

2. Leibniz 1998/1714: 260.

3. Leibniz 1998/1714: 260, opening the same paragraph from which the preceding quotation is drawn.

4. Suppose a "monad" is a complex form of mental activity supervenient on or rather implemented in a physical system along the lines of contemporary functionalism. Then read Leibniz 1998/1714: 37ff. on "homuncular functionalism" and the later discussion in Lycan 1987, followed by Lycan, "Consciousness as Internal Monitoring," in Block et al. 1997. Recall that Leibniz, the great visionary of formal thinking according to Husserl, was probably the first thinker to envision explicitly what we today call a computing machine.

5. A complementary anthology addressing philosophy of mind more broadly is forthcoming: *Phenomenology and Philosophy of Mind*, edited by the present author and Amie L. Thomasson.

6. Siewert (1998) argues at length from the case of blindsight for the need to recognize and not neglect the phenomenal features of consciousness in vision (see chaps. 3–5). I follow the same basic line of argument initially (do see Siewert's detailed discussion). However, Siewert argues "against ways of identifying consciousness with a kind of mentally self-reflexive feature" (197, see chap. 6 for details, noting 197). I concur that what defines consciousness is no form of higher-order thought or perception directed toward the conscious state (so noted in my 1986 and 1989: chap. 2). But I go on to seek an analysis of inner awareness that avoids those problems. Here I restrict inner awareness, though, to certain forms of consciousness. Stay tuned.

7. See the account of blindsight in Martha J. Farah, "Visual Perception and Visual Awareness after Brain Damage: A Tutorial Overview," in Block et al. 1997.

8. Siewert (1998: chap. 3) distinguishes a variety of cases of sight including full sight, amblyopic (partially atrophied) sight, empirically observed limited blindsight, and idealized full blindsight (as good as normal conscious vision).

9. See Smith 2002a.

10. Empathy is precisely projecting oneself from one's own first-person perspective into another's position as subject, comprehending the other's experience as if one's own. See Smith 1989: chap. 3.

11. Dennett (1991) introduced the useful term "heterophenomenology," albeit with the ulterior motive of gutting the concept of consciousness and undermining "autophenomenology," or phenomenology as I am practicing it.

12. See Freud 1965/1933.

13. The relations among forms of experience and forms of language are discussed at length in Smith and McIntyre 1982. The principles behind this method of phenomenological analysis are detailed there and carried over into Smith 1989, and into the present discussion.

14. See Smith 2002b.

15. As Brentano's students went to work on intentionality, two approaches developed: the "object" and the "content" approaches to intentionality, as appraised in Smith and McIntyre 1982. Meinong championed the object approach, while Husserl championed the content approach. On the object approach, all distinctions in intentionality are distinctions in the object "intended." On the content approach, there are distinctions in *content* as opposed to object, where the content defines or reflects the way the object is intended. Thus Husserl characterized the ideal content or sense (*Sinn*) of an act as "the object as intended."

16. Indexical intentionality is a complex story. See Smith 1989.

17. That perception is a form of indexical awareness, perhaps the paradigm form, is discussed at length in Smith 1989: chap. 1, and observed in Smith and McIntyre 1982: chap. 7.

18. This approach already has difficulties with intentionality: see Smith 1999. Different issues are raised here.

19. To be true to Brentano (1995/1874), we should note that for Brentano judgment is not necessarily of a subject-predicate form (142), as we here assume in line with today's logic. For Brentano "judgment" is an affirmation or denial (of existence), rather than a judgment that such and such is the case.

20. For the present interpretation, see Smith and McIntyre 1982: chap. 3, and (more briefly) McIntyre and Smith 1989. In Husserl, *Logical Investigations* (1970/1900–1), the distinction is drawn between the "quality" and "matter" of an act; in Husserl, *Ideas* I (1963/1913), the distinction is drawn between the "thetic character" and "*Sinn*" within the noema of an act.

21. In Searle 1983.

22. This tendency Siewert calls "the 'consciousness-of' trap" (1998: 194).

23. Compare Perry's recent account of "reflexive content" (2001: chap. 6).

24. For an illuminating discussion of the origin of the text (compiled in 1917 by Edith Stein from Husserl's lecture notes, then edited in 1928 with little change by Martin Heidegger), and an overview of Husserl's analysis and the difficulties of interpreting Husserl's evolving account of the "constitution" of time consciousness, see Brough's Translator's Introduction to Husserl (1991/1893–1917). A careful and succinct reconstruction of Husserl's analysis of time consciousness is that in Miller 1984.

25. Husserl's concept of "constitution" is central to his doctrine of "transcendental idealism," my take on which is elaborated in Smith 1995: 372ff.

26. A rich phenomenological study of related issues is that in Zahavi 1999. Zahavi carefully distinguishes aspects of the structure of self-awareness, including inner awareness and temporality.

27. The modal interpretation of "consciously" is a specific approach to the logic of the adverb "consciously." If the clause "I see this frog," or "I see that this is a frog," has a relational predicative form, the intentional verb "see" being a transitive verb (as Rosenthal and Dretske assumed), then "consciously" in the sentence "Consciously I see this frog" seems to take an adverbial form, modifying the verb "see." By grammatical analogy, the character of consciousness might be called an *adverbial* character, as Amie Thomasson (2000) has observed in addressing Brentano. I think that is correct, and that is the first principle to extract from Brentano. Thomasson contrasts such adverbial analyses (including mine) with models of higher-order thought or perception.

28. The details here were laid out in Smith 1989.

29. The interdependence of the various indexical contents – "this [object]," "here," "now," "I," "this very experience" – is studied at length in Smith 1989.

30. Searle (1992) holds that all mental states are conscious and that what some have called unconscious intentional states are really neural dispositions to produce conscious mental activities. Although I do not here directly address Searle's view, the delineation of levels of consciousness cuts against the sharp line Searle draws between conscious states and nonmental neural states.

31. The following scheme of four levels, and three biases, was drafted in my discussions with Chuck Dement around 1993. We were seeking a formal system of intentionality that would be framed ultimately by a formal ontology along the lines sketched in my essay "Being and Basis" (1997). A specific aim of the scheme was to place indexical modes of awareness in a larger formal phenomenology, assuming an account of indexical awareness like that in my book *The Circle of Acquaintance* (1989).

32. Johannes Brandl has made a similar point, in conversation and in a colloquium on reduction and levels of intentionality presented in spring 1999 at the University of California, Irvine.

33. See Smith 1989: chap. 1.

34. Searle (1983) characterizes "direction of fit" as one distinguishing feature of intentionality. For Searle, in thought or perception the direction of fit is from mind to world, as a state of thinking or perceiving fits the state of the world if the thought or perception is *true*; but in intention or action the

direction of fit is from world to mind, as a state of the world fits the intention in or behind an action if the intention is *satisfied*. The present conception of "bias" is somewhat like the notion of direction-of-fit, but distinguishes three forms of orientation or direction, separating thinking from perceiving and both from intention or volition in action. (I use the term "volition" where Searle uses "intention in action.") In D. W. Smith (1997) *bias* is cast as a basic mode of being: an entity is either receptive or static or active. Accordingly, in particular, a mental state is either receptive as in perception, or static with respect to environment as in thinking, or active as in volition *cum* action.

35. Systematics in biology is the science of the diversity of life forms. Here I use the term in an extended sense. Compare my "Basic Categories" in Chapter 8.

References

Block, Ned, Owen Flanagan, and Güven Güzeldere, eds. 1997. *The Nature of Consciousness: Philosophical Debates.* Cambridge, Mass.: MIT Press.

Brentano, Franz. 1995. *Psychology from an Empirical Standpoint.* Edited by Oskar Kraus. English edition edited by Linda L. McAlister, with a new introduction by Peter Simons. Translated by Antos C. Cancurello, D. B. Terrell, and Linda L. McAlister. London: Routledge. German original, 1874.

Chalmers, David J. 1996. *The Conscious Mind: In Search of a Fundamental Theory.* Oxford: Oxford University Press.

Dennett, Daniel C. 1991. *Consciousness Explained.* Boston: Little, Brown.

Descartes, René. 1988. *The Philosophical Writings of Descartes.* Vol. II. Translated by John Cottingham, Robert Stoothoff, and Dugald Murdoch. Cambridge: Cambridge University Press.

Freud, Sigmund. 1965. *New Introductory Lectures on Psycho-Analysis.* New York: W. W. Norton. German original, 1933.

Husserl, Edmund. 1963. *Ideas.* Translated by W. R. Boyce Gibson, from the German original of 1913. New York: Collier Books. Originally titled *Ideas Pertaining to a Pure Phenomenology and to a Phenomenological Philosophy, First Book.*

1970. *Logical Investigations.* Vols. 1 and 2. Translated by J. N. Findlay from the revised, second German edition. London: Routledge and Kegan Paul. New edition, edited with an introduction by Dermot Moran, and with a preface by Michael Dummett. London: Routledge, 2001. German original, 1900–1, revised 1913 (Prolegomena and Investigations I–V), 1920 (Investigation VI).

1991. *On the Phenomenology of the Consciousness of Internal Time-Consciousness (1893–1917).* Lectures from 1893 to 1917. Translated by John Barnett Brough. Boston: Kluwer Academic Publishers. From German text edited and organized by Edith Stein, then slightly edited under his name by Martin Heidegger for the 1928 text from which the translation was prepared. This edition includes further texts gathered by Brough.

Leibniz, G. W. 1998. "Principles of Nature and Grace, Based on Reason." 1714. In G. W. Leibniz, *Philosophical Texts.* Translated and edited by R. S. Woolhouse and Richard Francks. Oxford: Oxford University Press. Texts from 1686 to 1714.

Locke, John. 1975. *An Essay concerning Human Understanding*. Edited by Peter H. Nidditch. Oxford: Oxford University Press. Original, 1694.

Lycan, William G. 1987. *Consciousness*. Cambridge, Mass.: MIT Press.

McIntyre, Ronald, and David Woodruff Smith. 1989. "Theory of Intentionality." In J. N. Mohanty and William McKenna, eds., *Husserl's Phenomenology: A Textbook*, pp. 147–79. Washington, D.C.: University Press of America; Pittsburgh: Center for Advanced Research in Phenomenology.

Miller, Izchak. 1984. *Husserl, Perception and Temporal Awareness*. Cambridge, Mass.: MIT Press.

Perry, John. 2001. *Knowledge, Possibility, and Consciousness*. Cambridge, Mass.: MIT Press.

Petitot, Jean, Francisco J. Varela, Bernard Pachoud, and Jean-Michel Roy, eds. 1999. *Naturalizing Phenomenology: Issues in Contemporary Phenomenology and Cognitive Science*. Stanford, Calif.: Stanford University Press in collaboration with Cambridge University Press.

Sartre, Jean-Paul. 1963. *Being and Nothingness*. Translated by Hazel Barnes. New York: Washington Square Press. French original, 1943.

Searle, John R. 1983. *Intentionality*. Cambridge: Cambridge University Press.

    1992. *The Rediscovery of the Mind*. Cambridge, Mass.: MIT Press.

Siewert, Charles. 1998. *The Significance of Consciousness*. Princeton, N. J.: Princeton University Press.

Smith, David Woodruff. 1986. "The Structure of (Self-)Consciousness." *Topoi* 5 (2): 149–56.

    1989. *The Circle of Acquaintance*. Boston: Kluwer Academic Publishers.

    1995. "Mind and Body." In Barry Smith and David Woodruff Smith, eds., *The Cambridge Companion to Husserl*. Cambridge: Cambridge University Press. pp. 323–93.

    1997. "Being and Basis." Unpublished manuscript, extending my presentation under the same title at the Second European Congress of Analytic Philosophy, Leeds, England, 5–7 August 1996.

    1999. "Intentionality Naturalized?" In Jean Petitot, Francisco J. Varela, Bernard Pachoud, and Jean-Michel Roy, eds., *Naturalizing Phenomenology: Contemporary Phenomenology and Cognitive Science*, pp. 83–110. Stanford, Calif.: Stanford University Press in collaboration with Cambridge University Press.

    2002a. "What Is 'Logical' in Husserl's *Logical Investigations?* The Copenhagen Interpretation." In Dan Zahavi and Frederik Stjernfelt, eds., *100 Years of Phenomenology: Husserl's Logical Investigations Revisited*, pp. 51–65. Boston: Kluwer Academic Publishers.

    2002b. "Intentionality and Picturing: Early Husserl *vis-à-vis* Early Wittgenstein." *Southern Journal of Philosophy* 40 suppl.: 153–80.

Smith, David Woodruff, and Ronald McIntyre. 1982. *Husserl and Intentionality*. Boston: D. Reidel.

Thomasson, Amie L. 2000. "After Brentano: A One-Level Theory of Consciousness." *European Journal of Philosophy* 8 (2): 190–209.

Zahavi, Dan. 1999. *Self-Awareness and Alterity: A Phenomenological Investigation*. Evanston, Ill.: Northwestern University Press.

4

Consciousness in Action

Abstract. A phenomenology of action is outlined, analyzing the structure of volition, kinesthesis, and perception in the experience of action. These three forms of intentionality are integrated in a typical case of conscious volitional action. In their integration we find the structure of our everyday experience of embodiment in action. The intentionality of action is contrasted with that of thought and perception in regard to the role of the body, and the relations between an action, the experience of acting, and the context of the action are specified.

Segue. In "The Cogito circa A.D. 2000" we argued that the experience of consciousness includes an inner awareness of the act of consciousness (*sans* dualism and *sans* incorrigibility). In "Return to Consciousness" we analyzed the form of inner awareness in detail. Phenomenological analysis begins with our awareness of our own experience as lived. Often this phenomenological perspective, grounded in inner awareness, is assumed to lead to an isolation of mind from body. However, a careful phenomenology of action shows that this structure of consciousness already places consciousness in intimate relations to the "lived" body, so the "inner" is already tied to the "outer," the "subjective" to the "objective." Thus, consciousness is already embodied as we experience it in everyday action. Inversely, the body is itself experienced in conscious action. That is, consciousness is itself experienced as part of nature, beginning with one's own body.

An early version of this essay was read at the XVIIIth World Congress of Philosophy in Brighton, England, on 25 August 1988. I thank discussants on that occasion, including cosymposiasts Robert Gordon, Helen Lauer, and Jonathon Suzman; I thank Steve DeWitt, Ronald McIntyre, Martin Schwab, Peter Simons, and Peter Woodruff for discussions since; and I thank an anonymous referee for many detailed comments on the penultimate draft. The present edition is much longer than, and quite different from, what was read at Brighton.

The Phenomenological Problem of Action

As phenomenology describes human experience, the phenomenology of action describes the experience of *acting*, or *doing* something, especially doing something consciously and intentionally, or volitionally. Acting typically involves moving or using one's body by the *volition* to do such-and-such. But what is the structure of an *experience* of acting? That is the phenomenological problem of action.

Here I should like to present a blueprint for a phenomenology of action, sketching a theory of the intentionality of action – as opposed to thought and perception, which (at least on the face of it) do not involve using one's body. I shall assume the broad outlines of a "content" theory of intentionality, according to which an intentional state or "act" is directed toward an object via a content, where the content of the act prescribes, or is satisfied by, the object of the act (if the act is successful). In the case of thinking the content will be a proposition (e.g., "that 1970 was a good year for Bordeaux"), whereas in visual perception the content will be a visual image or percept (usually informed by conceptualization, as in seeing "this green frog here in the grass"). But in the case of action, we shall find quite different phenomenological structures, or contents, as we sort through the issue raised by human action for a general theory of intentionality.[1]

I do not propose to offer, as would some theories of action, an account of necessary and sufficient conditions for being an action. I am not sure there are any (given the plasticity of our concept of action). But in any event I want here to focus on a certain paradigm of *conscious, intentional* action. And, as we shall see, there are variations in what we experience in different actions even of this kind.

One does something consciously when one does it with awareness of what one is doing, and one does something intentionally when one does it with intention, or volition. (A distinction between intention and volition, to be drawn later, need not concern us at this time.) One may do something consciously but unintentionally (nonvolitionally). For instance, when I drive over the curb, I am driving the car consciously, and intentionally, but it is not part of my intention, or volition, to drive over the curb. Again, one may do something unconsciously yet intentionally (volitionally). For example, when I say something offensive to my friend, I may do so intentionally but unconsciously, motivated by envy. Our concern, however, will be those actions in which one does something both intentionally and consciously.

It is important to distinguish phenomenological problems from meta-physical problems. Because action involves bodily movement caused by volition, the traditional mind-body problem is just around the corner: in action, *how* is the *mental* state of willing related to (and *how* can it *cause*) the *bodily* state of moving and thereby doing such and such? That is a metaphysical problem. But another sort of mind-body problem arises in the phenomenological description of action. Conscious action involves awareness of one's volition and one's bodily movement. What is the relation between volition and bodily movement as experienced in conscious action? That is a phenomenological problem, formulated at the level of phenomenological description. By contrast, the tradi-tional mind-body problem is that of relating mental events as described in phenomenology or psychology with physical events as described in neurophysiology.[2]

In classical phenomenology, action is addressed more centrally by Merleau-Ponty than by Husserl or Heidegger. Merleau-Ponty is said to have undercut Cartesian dualism and averted the mind-body problem by describing the body in a way that does not admit of mind-body dis-tinctions. It would be better to see his account of "the body" as a phe-nomenological description of certain forms of intentionality, including bodily awareness in action and perception.[3] Similarly, in psychology, J. J. Gibson describes visual experience at a level of human rather than physical interaction, so that perceived objects are presented with proper-ties like "a stone to be stepped on."[4] At that level of description, objects in the "external" world are described in terms that overlap with descriptions of our "internal" world of experience. And that level of description – of one's experience and the world around one – is the level at which human actions are described in the phenomenology of action.

Intentionality in Action, Thought, and Perception

The phenomenological problem of action, I have assumed, is a problem of *intentionality* in the Brentano-Husserl sense: what is one conscious or aware of in performing an action?

But there are very different kinds of intentional activity. Some, such as thinking, are purely mental activities (realized in brain function), whereas others, such as action (and arguably perception), involve use of the body. These different kinds of "acts" have very different kinds of intentionality. Thus, I have a thought "about" Russell, a perception "of" the dog, and an intention or volition "to" hit the bull's eye. These different phenomena are all said to be intentional or "directed" toward something: thinking, because it is "about" something; perception, because it is "of" something; intention or volition, because it is "to" do something. Bodily actions, too, are said to be "intentional." But is the intentionality of action a special case of intentionality in the Brentano-Husserl sense? Not quite. An action is intentional in the everyday sense if and only if it is done with and through the volition or intention to act in a certain way, and a volition or intention is intentional in the Brentano-Husserl sense, as it is a willing or intending "of" the action, or "to" act thus and so. In this way, the everyday sense of "intentional" may be defined in terms of the Brentano-Husserl sense. Alternatively, the Brentano-Husserl sense may be extended (from purely mental acts) to actions, by saying an *action* is intentional insofar as the constituent volition (through which it is performed) is intentional.[5]

The classical phenomenologists took differing views of intentionality in part because of their diverging attitudes toward action and the body. Husserl focused on the intentionality of thought and perception as cognitive apprehension of objects in the world, whereas Merleau-Ponty focused on the intentionality of perception as essentially involving meaningful embodied behavior, or action. It is said that Husserl privileged cognitive representation in thought and perception divorced from action in the world, whereas Heidegger held that meaningful behavior (informed by culturally – or historically – defined practices) is more fundamental than thought or perception, and Merleau-Ponty (with less emphasis on praxis) urged that perception and bodily movement go hand in hand ("the theory of the body is already a theory of perception"). But these familiar oppositions blur what is at issue in the intentionality of action.

All intentional human activities – actions, thoughts, perceptions – take place *in the world* (where else?) and are performed by an embodied human subject. The issues should be: what are these *different kinds* of intentionality, in action, thought, and perception? And in what ways do they *depend* on each other – *is* one kind most fundamental, or *does* each depend in different ways on certain of the others? Once these questions are posed clearly, the familiar postures attributed to Husserl, Heidegger, and Merleau-Ponty appear wrongheaded: there are different kinds of dependence among these different kinds of intentionality.

Different Forms of Awareness in Action

What is one *aware of* in performing an action? Awareness comes in several things in different ways, in different forms of intentional activity.

Consider a relatively simple action: my picking up a shovel to dig in
the garden. We should distinguish the following forms of awareness,
or structures of experience, involved in my performing this conscious
action. First, within *action* itself, there is my volitional awareness of my
action, that is, my *volition* to so act: I am consciously willing to pick up
this shovel, that is, willing that I now pick it up (in a certain manner,
grasping the handle with my hands and lifting it with my arms). Second, accompanying my action is my *kinesthetic* awareness of my bodily
movement as I pick up the shovel. Third, there is my concurrent *perception* of certain things in my environment, especially my seeing and
touching the shovel I am picking up, and also my seeing the shovel
rising in my hands (the intended effect of my action). Thus, *my experience of acting* consists in my conscious volition, which occurs against the
background of my relevant kinesthetic and perceptual experiences. And
these three forms of awareness make up my *overall experience* in acting.
Importantly, although kinesthesis and perception accompany and guide
my volition, I am attending to what I am willing and thereby doing –
picking up the shovel – rather than to what I see or what I feel of my bodily
movement.[6]

Some philosophers balk at the claim that action involves a distinguish-
able element of willing and that we are aware of our willing in acting.
However, a striking experiment dissolves that skepticism. Neurologist
Wilder Penfield inserted an electrode into the motor cortex of the patient's brain, which caused the patient's hand to move. The patient, fully
conscious, declared, "I didn't do that. You did."[7] The point being that we
normally do experience our effectively *willing* our hands to move. The
phenomenology of action should explicitly recognize this element in the
experience of acting. To experience willing, however, is not to have something like a "feeling" of willing, a sensation of willing, like a churning in
one's stomach or, indeed, a kinesthetic sensation of moving. Consciously
willing is not sensation of any kind; it is willing.

The three forms of awareness noted in a typical action – volition, kinesthesis, and perception – are interdependent in that they presuppose one
another in certain ways. Thus, I normally cannot perform the conscious
action of picking up the shovel, by my volition to pick it up, without having
a kinesthetic awareness of my movement and a perceptual awareness of
the shovel and its movement in my hands. Nor can I see the shovel without turning my eyes toward it, or feel it without reaching my hands out
to touch it. These different forms of experience are thus interwoven as
I perform the conscious action. The action itself, as it unfolds, is guided

by the integration of these three forms of awareness of what I am doing. (Of course, there are many variations on this simple form of action, some of which we note later, but let us focus on the structures of experience involved in this simple paradigm.)

By hypothesis, these three forms of awareness involved in my action are all mental events. Now, what makes a mental event conscious is an "inner awareness" of it as it transpires. In particular, as I will that I now pick up this shovel, I am aware of my so willing. And, as I feel my body moving, I am aware of this kinesthetic sensation of my body. Again, as I see this shovel rise in my hands, I am aware of my visual experience of the shovel. However, this awareness of my intentional activity – my willing, sensing, or seeing – is not an *attentive* awareness, as my attention lies in what I am doing (raising the shovel) rather than in my awareness of my doing it. Furthermore, this awareness is not a distinct, second-order judgment or observation about the activity. It is rather a *prereflective* awareness of the activity, a *reflexive* awareness that is internal to the activity itself and makes it a conscious mental activity. The structure of my awareness in my volition, then, may be expressed in a phenomenological description of the form:

In this volition I am now willing (myself) to pick up that shovel, that is, willing that I pick up that shovel,

where "I am now willing that I pick up that shovel" articulates my volitional awareness-of-my-action and "in this volition" expresses my reflexive awareness-of-my-volition. A similar reflexive structure will be part of my kinesthesis and again of my perception. (A useful analogy with inner awareness is the feature of promising described in speech-act theory, where I say, "[Hereby, that is, in this very speech act] I promise to return the car by noon.")[8]

Thus, my overall experience in picking up the shovel is formed from three forms of conscious intentional activity:

1. My volition to pick up this shovel, and therein my inner awareness of my so willing.
2. My kinesthesis of my bodily movement, and therein my inner awareness of my kinesthesis.
3. My visual and tactile perceptions of the shovel's movement, and therein my inner awareness of my perceptions.

The *phenomenological structure* of these forms of awareness-*cum*-inner-awareness may be expressed in the following forms of phenomenological description:

1. In this volition I am now willing that I pick up – grasp with my hands and with my arms lift – this shovel that I am now feeling in my hands.
2. In this kinesthesis I am now sensing my arms moving upward as if lifting something and my hands closed as if around a shovel handle.
3a. In this visual perception I am now seeing this shovel in my hands and rising with my arms.
3b. In this tactile perception I am now feeling this shovel's handle in my hands.

It is in virtue of the reflexive awarenesses expressed by "in this volition," "in this kinesthesis," and "in this visual perception" that my volition, kinesthesis, and perception are conscious. (However, I am in no way attending to my willing; I am attending to what I am willing and thereby doing, namely, lifting the shovel. Nor am I in any way attending to my kinesthetic sensations while lifting the shovel, although I might attend to these sensations if concerned with my technique of lifting, say, if I am concerned not to strain my back. Nor am I attending to my perceptions, though I am attending to what I see of what I am doing, namely, lifting the shovel.)

My action itself is conscious insofar as I am aware of performing it. But in what ways am I aware of my action per se, my picking up the shovel? In light of the preceding, my *awareness of my action* consists in:

My willing my picking up the shovel.
My kinesthetic sensation of my bodily movement.
My visual and tactile perception of the movement of the shovel in my hands and arms.
My inner awareness of my willing, my kinesthetic sensing, and my perceiving.

Thus, I am aware of my action not only in my experience of acting per se, which consists in my conscious volition of my picking up the shovel, but also in my kinesthetic sensations of my movement and my perceptions of the action.

As indicated earlier, my experience of acting is an experience of consciously executing the action. As I perform the action, I do not experience (I have no awareness of) any difference between my action of successfully raising the shovel and my experience of acting. I simply experience my volition as effective – unless, through perception or kinesthesis, I become aware of failing to do what I am willing, say, if the shovel slips out of my hands. (Similarly, in perception I do not experience any difference between my successfully seeing an object before me and my experience of seeming to see such a thing – unless something happens to shake my normal faith in my perception.)

Moreover, my experience of acting involves an *experience of causation*.[9] To flesh out the phenomenological structure of my volition:

> In this volition I am now willing that I pick-up-this-shovel *by so willing*, that is, that I pick up this shovel and this volition cause my bodily movement in picking up this shovel.

Thus, the content of my volition prescribes – its conditions of satisfaction are – not only that my body move in a certain way but also that its movement be caused by that very volition and, furthermore, that this movement cause the shovel's upward movement.[10] I experience this causation precisely insofar as I experience the volition, and collaboratively the kinesthesis and perception that attest to its causal efficacy, and these I experience in virtue of the inner awareness that makes them conscious. In this way I experience the whole causal process of volition bringing about bodily movement bringing about shovel movement.

Notice that what I am willing – my action of picking up the shovel – incorporates not only my bodily movement (my hands grasping and my arms lifting) caused by my volition but also the shovel's movement caused by my bodily movement. In a tai chi exercise, however, what I am willing is only a certain bodily movement and, through it, a certain mental state of concentration; I am not willing any physical activity beyond my bodily movement. In tai chi, of course, I pay more attention to my kinesthetic sensation of my movement and very little to my surroundings or my perception of my environs. By contrast, in a ballet exercise I am willing my bodily movement and watching my movement in the mirror, and then I may attend more to my visual image than my kinesthetic image of my movement.

Action, Experience, and Context

Having picked up the shovel, I am digging with it, digging a hole, where I shall plant a rosebush, here in the garden. What are the boundaries of my action of digging and my experience of digging; that is, what are the structural parts of the *action* itself, the *experiences* involved in acting, and the *context* of action?

My action of digging consists in the complex process or event that involves: (1) the digging itself, that is, my bodily movement plus the movement of the shovel in my hands and the earth in the shovel head; (2) my volition in digging, that is, my willing "that I dig here by moving this shovel [in a certain way] with my hands and arms" (which is accompanied by the relevant kinesthetic and perceptual awarenesses); (3) the causal relation between my volition and my digging movement; and (4) the intentional relation between the volition and the action, that is, my volition's willing, or being a volition "of," my digging. The relationship between movement and volition in action is thus quite involved. Not only is the movement both caused and willed by the volition, but, as noted, the causation is part of what is willed and part of what is experienced in willing (and in the kinesthesis and perception of moving).

My *experience of acting* is that part of the action in which I experience the action. Thus, my experience of digging consists in my conscious volition of digging, in the company of kinesthetic and perceptual awarenesses, and my volition includes willing my movement with a reflexive awareness of my so willing. As I perform the action, my digging just *is* – in effect, for me – my experience of digging, my consciously executing my volition of digging. From my first-person, in-action perspective, there is no gap between my action and my volition of acting.

When I am trying to do something very precise, as in executing a golf swing, however, I may well not move my body in just the way I am trying, or willing, to move. In kinesthetic awareness of my movement, or perceptual awareness of the club's or ball's movement, I may then notice mistakes or inefficacies in my action. In that case I am aware, in performing my action, of the difference between what I am willing and what I am doing. I become aware that I am trying as opposed to doing, willing but not succeeding.

Indeed, the possibility of error in action requires that we distinguish the experience of acting from the action itself, which includes my actual bodily movement. For, in theory at least, if I were hallucinating, I might have an experience of so acting – consciously so willing in the company of appropriate kinesthetic and visual experiences – even if I move not a muscle and so perform no such action. Normally, however, when I want to move, I move: my volition is effective.

Although my bodily movement in digging is distinct from my volition of digging, both are parts of the action. On the present account, however, my kinesthetic awareness of my digging motion is not part of the action per se but is a kind of perception of my bodily movement. (Sometimes this is called "proprioception," meaning perception of self, that is, one's bodily self; the perceptual organ consists of neural networks that transmit information about muscle activity rather than, say, ambient light or heat.) Similarly, my perception of the shovel and its moving earth is not part of the action but is a perception of what the action is achieving. My accompanying kinesthetic and perceptual experiences are, thus, part of the context of my action, part of the "background" phenomena on which the action depends.

Furthermore, the kinesthetic and perceptual experiences are not part of my experience-of-acting per se, which consists in my volition of digging. Nonetheless, these three forms of experience are integrally related, as they are interdependent in various ways, and collectively they form my overall experience in acting. Phenomenologically, my bodily awareness and even my perception of the shovel's moving earth seem inextricable from my experience of digging. Indeed, the perception requires moving my head and eyes in the right direction, and that movement is an integral part of my action of digging insofar as I must move my head to look where I am moving the shovel. The neurophysiology of action would support this integration, mapping the complex systems of feedback among motor control (volition), muscle feedback (kinesthesis), and visual intake (perception). Reflecting this structure of interdependence, our phenomenological description of my experience of acting sets my volition of digging against the background of my accompanying kinesthesis and perception. In this way, kinesthesis and perception form part of the proximal context of the volitional experience of acting.

So far, we have not distinguished volition and intention. But suppose that I have for weeks intended to plant a rosebush in a certain corner of the garden. Today, I am digging here in the garden, with the intention of planting a rosebush there. My volition in digging is an event that is a constituent part of my current experience (and part of my action) of digging, but my intention is rather a state that is part of the background of my current experience of digging. Thus, behind my volition in digging, presupposed by my volition, is my intention to plant a rosebush there. The willing is contemporaneous with the digging, but the intending stretches further back in time. The state of intending is not itself causally effective, producing the action; the event of willing is. Thus, the volition carries out the prior intention it presupposes.[11] The intention itself, moreover, issues from my desire to put in a rosebush, coupled with my beliefs about how to do this.[12] And my volition is further shaped by my emotions and my mood, of the moment. (The cry "There's a fire in the garage!" would immediately change the course of my volition and my action.)

The intention, desires, emotions, moods, and beliefs behind my action are part of the context of the action, specifically, the background of states and events on which the action depends. In case the action is part of a larger project, as in my digging while revitalizing our garden, my knowledge of that project and how the action fits into it – however vague that knowledge may be – belongs to the background of the action. Also part of the background is my know-how, including my manual skills exercised in digging with a shovel. Still another part of the background on which the action depends is my implicit cultural understanding, for example, my understanding of the "significance" of digging in our culture, something that may not even be explicable in terms of representational knowledge. Still a different part of the background of the action is the structure of cultural practices on which the action draws, for example, maintaining gardens.[13]

As explicated earlier, the structure of my action, my experience in acting, and the relevant context of my action may be schematized as follows, the main features depicted in the accompanying Figure 4.1:

The *action* = (body movement + shovel-and-earth movement, volition, causal relation, intentional relation).
The *experience* of acting = (volition).
The *context* of the action = (action, psychic context, physical context, social context).

where:

The *psychic context* of the action = (volition, mental-event context, mental-state context).
The *mental-event context* of the action = (volition, kinesthesis, perception).
The *mental-state context* of the action = (volition, intention, desires, emotions, moods, beliefs about means, motor skills, knowledge of project, know-how involved in project, understanding of cultural practices involved).
The *physical context* of the action = (agent's environs, causal relations between the action and other events).
The *social context* of the action = (action, social groups of agent, cultural practices involved, cultural history).

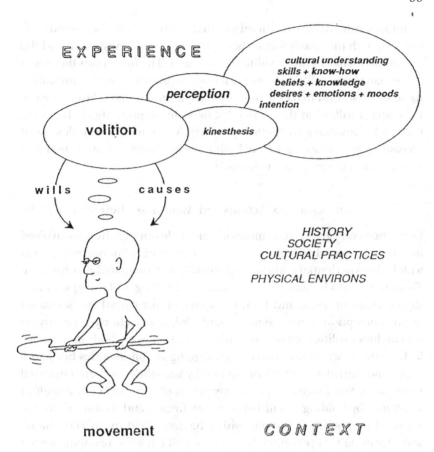

FIGURE 4.1. The structure of action, experience, and context.

The *intentional* states or events within the context of an action include: the *mental events* of volition, kinesthesis, and perception; and the *mental states* of intention, desire, emotion, mood, belief-about-means, and knowledge-about-project. *Nonintentional* states within the context include know-how and cultural understanding. Within the context of action, one's consciousness in acting lies in three events: conscious volition (with reflexive awareness), conscious kinesthesis (with reflexive awareness), and conscious perception (with reflexive awareness). The volition is part of the action proper, constituting the *experience of acting,* whereas the kinesthesis and perception are part of the context of action.

Importantly, we have included within the action itself, as opposed to its context, both the *causal relation* between volition and movement and the *intentional relation* between volition and action. The first makes the action more than mere bodily movement (it is appropriately caused by my willing it), and the second makes it "intentional" in the Brentano-Husserl sense (it is what I will, what the content of my volition prescribes). Together, these relations make the action properly "volitional": a complex event consisting in movements of my body and my shovel in both causal and intentional relations to my volition.[14]

More Complex Actions and Awareness Thereof

Our actions vary greatly in complexity, and so do our experiences involved in acting. Extremely simple actions may consist of volitional movements with little experiential content and virtually no contextual involvement. Turning my eyes leftward, for instance, I am willing and doing no more than moving my eyes, and I am not aware of any kinesthetic sensation or any perception of my eye movement. Doing a tai chi exercise, my action includes volition guided by paying attention to my body movements in kinesthetic sensation, whereas my digging action includes body-and-shovel movement and a volition guided by kinesthesis and even external perception. Still more complex is my action of securing a loose shelf in a cabinet, by holding a nail between the fingers and thumb of my left hand, while pounding the nail with a hammer held in my right hand, while focusing my eyes on the head of the nail (!), while propping myself up on my right knee, while visualizing the intended angle of one board vis-à-vis another.

Or, consider the indescribable complexity in a ballerina's performing a dance solo like "The Dying Swan," her consciousness currently focused on a certain form of movement, itself part of a lengthy piece of choreography accompanied by the music and generating a certain aesthetic mood for her and her fellow dancers in the wings and the audience. How does the character of her action depend on its context, including the cultural significance of dance, the intentions of the choreographer, and her interpretation of the choreography? And how does the content of her volition in acting reach into these cultural formations? Such cases go far beyond the relatively simple cases we have described. Yet even a simple case of digging calls for further analysis along these lines (although we cannot pursue the analysis here).

The complexity of one's experience-of-acting depends not only on one's concerns, in what one is trying to do but also on the focus of one's attention. Thus, I may experience the same form of action in different ways. Consider a complex and skilled form of action, for example, hitting a serve in tennis. If all goes well, I am merely aware of hitting a spin serve to the backhand corner (anticipating already a crosscourt return to my backhand volley). But if I have not been serving well today, I may pay more attention to the "feel" of the stroke, that is, the pattern of kinesthetic sensations during the movement (am I dropping my elbow too low again?). My volition may then be more complex: instead of trying (willing) simply "to hit a spin to his backhand," I try "to hit my best spin serve by bringing my elbow up." Or, if I break a racket string on impact, I am suddenly aware – perceptually – of the impaired racket I am swinging through the ball, and I may suddenly refine my volition, trying "to hit out well through the ball with my broken-stringed racket." Whereas, when all went well, I was concentrating only on executing my volition "to hit a spin to the backhand." I was then aware of willing my performance of that familiar movement, in one fell swoop (!), but virtually unaware of the racket in my hand and virtually unaware of any kinesthetic sensations of my movement or visual or auditory perception of the racket meeting the ball.

Attention is, thus, a dimension of phenomenological structure we have not stressed in the prior analysis. In particular, each form of awareness in my experience of action carries a certain degree of attention distributed over its objects. It is in this respect that I am "more aware" of this or that in acting: of the *shovel* (in visual or tactile perception of it), of my *arms* (in kinesthetic awareness of their movement), or my *willing* (in reflexive awareness), or my *kinesthetic sensations* of my arms moving (in reflexive awareness), or of my *seeing* the shovel moving (in reflexive awareness). A more complete phenomenology of action would specify a measure of attention and its structural distribution in these forms of awareness.

There are, then, very different levels of complexity in action and in experiences of acting: from simply moving one's eyes or hands; to digging, moving the shovel through the earth with full attention on the parting earth; to hitting a twist serve, mindfully keeping one's elbow up to get full stretch and full power; to performing a dance solo. These different forms of action may invite very different approaches to the phenomenology of action and its intentionality, stressing the agent's larger project (cf. Sartre), or the cultural practices presupposed in acting (Heidegger), or the essential embodiment of intentional action (Merleau-Ponty), or simply the kinesthesis involved in moving (Husserl). Indeed, extremists might seek to reduce all intentional phenomena to one of these phenomena, for instance, reducing thought to modifications of social practice (an extreme reading of parts of Heidegger).[15] However, a sensitive theory of the intentionality of action must explicate all these aspects and complexities of action and the experience of acting.

Practical Intentionality

How does the *practical* intentionality of action, or volition in action, differ from the *cognitive* intentionality of thought and perception?

Consciously digging a hole with a shovel, willing one's digging, seeing or feeling the shovel's movement, sensing the movement of one's body in digging, thinking about digging a hole, thinking about oneself (say, while digging) – these are all intentional activities. But they have very different kinds of intentionality. In particular, the action of digging is a practical activity, not a cognitive activity like thinking or perceiving or, indeed, kinesthetically sensing one's movement. It is a practical activity insofar as it is one of acting, and it is not itself a cognitive activity of thinking or perceiving or conceiving (although it depends on such cognitive activities). Now, an action is intentional insofar as it is performed through a volition. And the volition itself is a practical intentional activity, as opposed to a cognitive intentional activity, because it is a willing rather than a thinking or perceiving (on which it may, however, depend). But what sets the practical intentionality of volition apart from the cognitive intentionality of thought, perception, and the like?

The difference begins with content. My volition of digging has a *volitional* content:

"In this volition I am now *willing* that I am digging this hole."

By contrast, my thinking of my digging has a *cogitative* content:

"In this thought I am now *thinking* that I am digging this hole."

Moreover, if Castañeda is right, the volitional content expressed here by "I am digging this hole" is of a different *logical* or *phenomenological form* than the cogitative or propositional content expressed by "I am digging this hole."[16] The content formulating what is willed, as opposed to thought, he calls a *practition*, as opposed to a *proposition*, the difference lying in the copula: while the relevant proposition has the form "I *am* digging" ("In this thought I think that I am digging"), the practicion has the form "I *to* dig" ("In this volition I will I [myself] to dig"), and the copula "to" signals a "causal openness" in the volition. Whether or not these core contents of thought and volition are different in form (we need not settle that matter here), the overall contents certainly differ in form.

Moreover, the activities that bear these contents differ in effect. A volitional content is, if you will, an *executive* content, in that "performing" a volition executes an action – if all goes according to form, that is, according to content. And this is what most fundamentally distinguishes the intentionality of volition from that of thought or perception; this is what makes it a practical intentionality.

Indeed, what most strikingly distinguishes volition from thought or perception is the causal efficacy of volition. Whereas I represent the world in thinking, I seek to affect the world in acting, by causing my bodily movement through my own volition. Thus, willing my digging causes my movement in digging and thereby executes my action: that is what makes the volition a practical intentional activity. By contrast, thinking that I am digging does not in itself cause my movement in digging; nor does kinesthetically sensing my movement cause my movement – quite the reverse. And not only does my volition cause my movement, but in willing I experience the causal relation between my willing and my movement (my awareness of causation being confirmed by my kinesthetic and perceptual awareness of my movement). In this way the ontology and phenomenology of action come together in causation.

Still, as different as practical and cognitive intentionality are, they share a familiar ontological structure: the "act-content-object" structure articulated by Husserl and others of the Brentano tradition. For the case of thought, this structure is outlined in the following principles:

1. An *act* of thinking has a certain *content* (a thought or proposition) that *represents* – semantically *prescribes*, or is *satisfied* by – a certain *object* (a state of affairs).
2. The *same object* can be represented in different ways, that is, in acts of thinking with *different contents*.
3. An act of thinking may have a content but *no object* (a false thought represents no actual state of affairs); or it may have an object that does not exactly fit the content.
4. The object of an act of thinking always *transcends* the content of the act, in that there are more aspects of the object than can be

represented in the content. (This applies not only to the state of affairs represented, or thought to obtain, but to the constituent objects thought "about.")

5. An act of thinking is *intentionally related* to an object if and only if the act has a certain content and that content *prescribes*, or is *satisfied* by, the object – that is, the intentional relation is the composition of the relations of entertaining and (successfully) prescribing.[17]

Similarly, for the case of action:

1. An *action* includes a bodily movement and a *volition* that has a certain *content* that prescribes a certain object, namely, the action itself.

2. The *same action* can be willed in different ways, that is, in volitions with *different volitional contents* (including contents that vary in complexity).

3. A volition may have a content but *no object* (where the volitional content is not satisfied, when the attempted action fails, as in missing the nail with the hammer, or when no movement ensues, as in paralysis); or a volition may produce a movement or effect that does not exactly fit the volitional content (say, in executing a ballet step).

4. The action willed (including the bodily movement) always *transcends* the content of the volition, in that there are more aspects of the action (e.g., bodily nuances, causal effects, and cultural significances) than can be represented in the content.

5. The *intentional relation* between a volition and an action consists in the volition's having a certain content and the action's satisfying the content.

Notice (again) that this intentional relation is distinct from the causal relation between the volition and the movement in the action (although the causal relationship is part of the conditions of satisfaction of the volition's content). That is why action is an "intentional," not merely a causal, phenomenon.

The principle of transcendence (4) is a point of special importance. The beauty of action in practice is that the content of my volition may be most minimal ("dig this hole," "hit a spin serve there"), yet my bodily movement may be exceedingly complex, the result of years of practice, and my actions may be informed with deep cultural significance (reflecting the role of gardens or games in our culture) – with little or none of these complexities being explicitly represented in the content of my volition itself. The writings of Heidegger and Merleau-Ponty, and Wittgenstein too, are rich in suggestions of this transcendence. However, we should not infer (as some of their interpreters are wont to infer) that there is no such thing as intentional content – or, indeed, consciousness – in action. In the (post)modern metaphor: the form of bodily movement in an action may be "compiled" from vast complexities of neural-network activities, and the form-of-action itself from a vast cultural network of interlocking forms of action, although the volition that consciously executes the action carries only the most minimal content. The job of phenomenology of action is to characterize such contents carefully, defining their intentional force and, in a larger context, their limits as well. (The larger context is the domain of "connectionisms," tracing neural, conceptual, and cultural networks.)

The Bodily "I" in Action

The *body* plays a central role in practical intentionality (as long as we are not gods or angels). Not only does action itself involve bodily movement, but the experience of acting involves an awareness of oneself as an acting, bodily "I."

In my action of digging, my body in fact plays three distinct roles: as *part*, *means*, and *end* of action. First, the movement of my body is part of the action. Second, this movement (caused by my volition) causes the intended effect of the shovel's dislodging earth in the garden. Third, the content of my volition prescribes a certain movement of my body (caused by the volition).

By contrast, the body plays no such essential roles in most cognitive intentional activities. Thoughts about one's body make intentional reference to one's body, but thinking does not make use of the body in the way that digging or running does. (I "use" my brain in thinking, but that is another matter.) To be sure, there are indirect ways in which thinking may involve bodily action, for instance, the ways in which thinking depends on the use of language, especially while thinking "out loud." But the role of the body in such thinking is very different from its role in action. Perception is the one cognitive intentional activity in which the body plays a role like that in action. For not only does vision involve a causal relation (barring hallucination) between one's environment, one's body (eyes), and one's visual experience; the content of the visual experience also prescribes the causal relation, and so makes intentional reference to the perceiver's body.[18] Moreover, one must (normally) turn one's head and eyes in the direction of an object in order to see it, whence the perception itself presupposes this simple action, which involves bodily awareness.

Now, in my action of digging, I both move and experience my bodily movement. Or, better, I experience myself moving. But I do not experience myself as a disembodied, or even disembodiable, "I" that wills the movement of "this body" and receives kinesthetic and visual images of the movement. Perhaps I experience myself as a purely mental "I" when I think: *cogito, ergo sum.* (Or do I? It was only through a strenuous course of argument that Descartes's *Meditations* led to the conclusion that "I" am a purely thinking being.) But when I dig and experience my digging, I experience myself as a moving, acting, bodily being, and this is essential to the structure of my experience of digging.

Thus, in the experience of acting, in the structure of volition, we find a certain element of self-awareness. In our phenomenological description of my experience of digging, this mode of self-awareness is indicated by the word "I":

> In this volition I am now willing that I am digging this hole with this shovel.

The volition itself occurs against the background of, and presupposes, kinesthetic and perceptual experiences whose structure we describe so:

> In this kinesthesis I am now sensing my hands closed as if around a shovel handle and my arms moving forward and upward as if in digging.

> In this visual experience I am now seeing this shovel in my hands moving this earth.

What is the force of this content "I" in the experience of acting – in the volition and in the accompanying kinesthesis and perception?

The word "I" occurs twice in our description of the volition.[19] Thus, the content "I am digging . . ." prescribes the action willed. Clearly, this content presupposes that "I," the agent, am an embodied being capable of moving in a digging way, and moving volitionally. Now, the reflexive content "In this volition I am now willing . . ." characterizes my inner awareness of my willing. But while willing does not itself include bodily movement, this content nonetheless presupposes that "I," the willing subject, am capable of action and thus embodied, because what "I" will is that "I" (the same being) am digging. Thus, the intentional force of the content "I," in both occurrences, is that of prescribing an *embodied agent.* This prescription is further supported by the content "I" in the accompanying kinesthesis and perception. For the kinesthetic experience presents "my" bodily movement, movement of this same bodily "I." And the visual experience also presents "my" bodily movement, "me" moving, from a different perspective. So, again, the "I" so experienced is a bodily "I." I also experience myself as a subject of kinesthesis and perception, through the reflexive contents "in this kinesthesis I sense . . ." and "in this visual experience I see . . ." In my overall experience in action, then, I experience myself as a being that sees things, feels bodily movement, and wills and performs actions.

In acting, we might say, I experience myself as body, I experience "I" as "my body," "me-body." Indeed, in the idiom of an older English, we are "bodies" as much as "persons" or "selves." We still say, "Is anybody home?" And, the old lyric says, "When a body meets a body, comin' through the rye." In today's English, however, we tend to associate more physiology and physics with the word "body."

Indeed, the object of this complex experience of oneself in acting transcends what the content "I" prescribes of it. (This accords with the principle of transcendence noted earlier, for practical as for cognitive intentionality.) In particular, what natural science has to say about this living-moving-willing "body" that "I" am is quite different from what my experience in action tells me about it. The phenomenology of action, body, and self is one story, their physics and physiology quite another.[20]

In the context of action theory, then, we need to address not the familiar, Cartesian mind-body problem (how can a mental event of volition cause a physical event of body movement?) but, instead, the *body-body problem*: what is the relation between the consciously-willing "body" and the physiological "body," between the body-as-it-appears-in-the-experience-of-acting and the body-in-itself-in-physical-nature?[21]

Notes

1. This kind of theory of intentionality is characterized in Smith 1989. Its locus classicus is in Husserl's work, according to the interpretation in Smith and McIntyre 1982; its history, beginning in Aristotle, is traced in Sajama and Kamppinen 1987. Where I speak of the "structure" of an experience of acting, I understand structure as "content" in that traditional sense. However, I wish to remain neutral here on the ontology of contents, as the results presented here can accommodate different conceptions of content.

2. In his modern classic "Mental Events" Davidson (1980) argued that these two levels of theory cannot be crossed by bridge laws relating the mental and physical (especially as cause and effect), although the two levels are ultimately about the same events albeit differently described: thus, Davidson's "anomalous monism." My point here is that the problem of action's being caused by volition already arises at the level of phenomenological or mentalistic psychology.

3. See Merleau-Ponty 1962: "Part One: The Body." Also, on "incarnate meaning" in Merleau-Ponty, see Dreyfus 1982. An elaboration of a partly Merleau-Pontian theory of action is found in Dreyfus and Wakefield 1988.

4. See Gibson 1979. A Merleau-Pontian reading of Gibson is developed in Blinder 1986.

5. Among recent action theorists, Hector-Neri Castañeda and John Searle assume the intentionality of action is a species of intentionality in the sense characterized by Brentano and Husserl. See Castañeda 1975 and Searle 1983. However, the issue of intentionality in the Brentano-Husserl sense is not explicit – so as to point toward a bona fide theory of intentionality for intention and action – in the earlier classics of action theory, for example, in Wittgenstein 1953, Anscombe 1959, Melden 1961, and Davidson's "Actions, Reasons, and Causes," in Davidson 1980.

6. The sense of body in action was elaborated by Husserl and Merleau-Ponty. See Husserl 1989 and Merleau-Ponty 1962. Cf. Russow 1988. What is at issue in "bodily intentionality," in the Merleau-Pontian claim that the body has a unique form of intentionality that is not reducible to the intentionality of thought? Does "bodily intentionality" mean *embodied* intentionality, say, in perception and action? Or, does it mean awareness *of* one's body, in perception and action? Or, does it mean the body itself *as* experienced in perception and action, the body at a certain level of intentional description?

7. Cf. Penfield 1975: 76. Searle (1983, pp. 89ff.) stresses the point made here, but he calls the volition "intention-in-action."

8. This reflexive form of "inner awareness" of an experience is explicated in Smith 1989: chap. 2; a shorter version is presented in Smith 1986. There I draw upon, and critique, Sartre's doctrine of "prereflective *cogito*" and Brentano's doctrine of secondary intentionality (that every mental act includes a consciousness of itself as well as a consciousness of its object). See Brentano 1973: 153ff. and Sartre 1966: introd. Importantly, I concur with Castañeda's analysis of the first-person form of the content of an intention, carrying the same form over into volition; thus, Castañeda would use the "quasi indicator" "he himself" to form a third-person attribution of the form of volition, as in "Smith intends/wills [himself] to pick up that* shovel," that is, "Smith intends/wills that he himself pick up that* shovel" (where "that*" is the quasi indicator that invokes Smith's own visual perspective on the shovel before him). See Castañeda 1975: chap. 6. One terminological difference: Castañeda says what is intended is not an action, and I say it is; but by "action" he means a type of event, or action-type, whereas I mean a particular event, or action-token.

9. Searle describes this "experience of causation" in Searle 1983: chap. 3. I essentially concur, observing similar details of content, except that I replace intention in action with volition. Cf. Castañeda's explication of what he calls the "internal causality of practical thinking" (1975: chap. 10).

10. The volition is *intentionally related* to, or successfully "of," an event – an action – if and only if the content of the volition (successfully) prescribes, or (inversely) is satisfied by, that event. In Searle's idiom, which I take to be equivalent, the action is the "condition(s) of satisfaction" of the volition's content (see Searle 1983: 80). The action, on the present analysis, comprises the occurrence of a bodily movement together with the volition's causing the movement and (as the case goes) the bodily movement's causing the shovel to move appropriately. For further elaboration of this notion of intentional relation and the role of content in intentional relations, see Smith 1989: introd. and chap. 4.

11. Searle distinguishes prior intention from intention in action (see Searle 1983: chap. 3) because he assumes the intention-in-action causes the action. I take it that by "intention in action" he means what I have called the volition in the action. However, Searle tends to use the word "volition" differently than I. First he talks of the "traditional category" of volition (p. 31), including desire as opposed to intention. Then, introducing "the experience of acting" as a term of art for the intention in action, he disparages "what some philosophers have called volitions or acts of willing" because our experiences of acting "are not acts at all" (pp. 88–89). I agree with his point, but I prefer the term "volition" or "willing" where he uses "intention in action" because "intending" sounds like a passive state rather than a causally efficacious event.

12. On Davidson's broadly Aristotelian model, the desire to achieve a certain end (or "pro-attitude" toward it) together with the belief that-doing-A-will-achieve-that-end normally tends to cause the action of doing A – on pain of weakness of will. See "Actions, Reasons and Causes" and "Weakness of Will," both in Davidson 1980. On the present model, the desire and belief may give rise to the intention to do A, but there is still slack (as philosophers from Plato to Davidson have felt): the intention is not sure to produce the action, as volition may not kick in. Thus, "weakness of will" is here weakness of intention. That is why volition must be distinguished from intention: it is volition, and not yet intention of desire-plus-belief, that is causally effective and issues in action. Michael Bratman has stressed the role of planning in action: see Bratman 1987. On the present model, planning would be part of the process of framing an intention in response to desire plus belief (although Bratman distinguishes planning theories from belief-desire theories of action). I would resist the assumption that all actions involve something like a deliberate and deliberative plan; if plans are to be an essential part of action, then the notion of plan must be watered down to something like the content of an intention (which is not necessarily the result of deliberate planning).

13. Skills belong to what Searle calls the "Background" of an intentional state, as distinct from the "Network" of beliefs associated with the content of the

intentional state. See Searle 1983: chap. 5. Heidegger stresses the background cultural practices on which everyday activities depend, their historical formation, and our implicit understanding of their significance. See Heidegger 1962: secs. III and V. Sartre sets every action within a larger, guiding "project"; see Sartre 1966: part 4. Developing his interpretation of Merleau-Ponty and Heidegger, Dreyfus stresses skills, formed by cultural practices, as virtually definitive of action; see Dreyfus and Wakefield 1988 and Dreyfus 1991: 55ff.

14. An intentional relation is the "semantic" relation that holds between an intentional experience, or its content, and its object. I have argued elsewhere against collapsing intentional relations with causal relations, in particular, in the analysis of perception. See Smith 1989: especially chap. 4. Similarly, I here oppose collapsing them in the case of action or, rather, volition. I read Searle as of like mind in his discussion of the experience of causation (see again Searle 1983: chaps. 3 and 4).

15. See Haugeland's (1982) quasi-behaviorist or pragmatist interpretation of Heidegger. Hubert Dreyfus, drawing alternately on Heidegger and Merleau-Ponty, has long argued against Husserl and cognitive science on grounds that they impute too much cognitive structure in situations where human beings simply act, drawing on nonintentional skills and extant human practices. See the flow of argument in Dreyfus 1991. The most detailed explication of his views relevant to my issues here is in Drefyus and Wakefield 1988. I believe the present account of action is consistent with their insights in that essay, although Dreyfus would disagree. The outstanding problem is to articulate – in still greater detail – the interdependence in action among conscious intentional events (volition, kinesthesis, perception, etc.), nonintentional states (skills, habits, etc.), bodily states (including neural network states), and social states (invoking cultural practices, etc.). It is essential, in any event, to detail carefully the limits of the contents of volition in various types of actions. But all this must await another occasion.

16. See Castañeda 1975: 28off.

17. These basic principles about intentionality are expounded in Smith and McIntyre 1982: chap. 1, and further explicated in following chapters.

18. Details of this intentional relationship are explored in Searle 1983: chap. 4 and Smith 1989: chap. 1.

19. The structure of self-awareness here described follows the pattern in Smith 1989: chap. 2. In the idiom of that work, the content "In this volition I am now willing . . ." is the modality of presentation in the volition, including the mode of inner awareness of oneself, and the content "I am digging . . ." is the mode of presentation of the action in the volition.

20. Accordingly, Husserl distinguished two conceptions of body, that of the "lived" or "living body" (*Leib*, from the verb *leben*, meaning "to live") and the "physical body" (*Körper*, from the Latin *corpus*, for "body"). See Husserl 1989: sec. 38 on *Leib* as "organ of will," and Husserl 1954: sec. 28ff. Merleau-Ponty (1962) followed Husserl's lead, distinguishing "phenomenal" and "physical" body.

21. Addressing eliminative materialism, Arthur Danto (1985) used the felicitous phrase "the body-body problem," independently.

References

Anscombe, G. E. M. 1959. *Intention.* Oxford: Basil Blackwell.

Blinder, David. 1986. "A New Look at Vision." *Topoi* 5: 137–48.

Bratman, Michael. 1987. *Intention, Plans, and Practical Reason.* Cambridge, Mass.: Harvard University Press.

Brentano, Franz. 1973. *Psychology from an Empirical Standpoint.* Translated by A. C. Rancurello, D. B. Terrell, and L. L. McAlister. London: Routledge and Kegan Paul. German original, 1874.

Castañeda, Hector-Neri. 1975. *Thinking and Doing.* Dordrecht: D. Reidel.

Danto, Arthur. 1985. "The Body-Body Problem." Paper presented as a University Lecture at Columbia University, 13 November.

Davidson, Donald. 1980. *Essays on Actions and Events.* Oxford: Oxford University Press.

Dreyfus, Hubert. 1982. "Husserl's Perceptual Noema." In Dreyfus, ed., *Husserl, Intentionality and Cognitive Science,* pp. 97–123. Cambridge, Mass.: MIT Press.

1991. *Being-in-the-World.* Cambridge, Mass.: MIT Press.

Dreyfus, Hubert, and Jerome Wakefield. 1988. "Action and the First Person." Unpublished manuscript, 25 February.

Gibson, James J. 1979. *The Ecological Approach to Visual Perception.* Boston: Houghton Mifflin.

Haugeland, John. 1982. "Heidegger on Being a Person." *Noûs* 16(1): 15–26.

Heidegger, Martin. 1962. *Being and Time.* Translated by John Macquarrie and Edward Robinson. New York: Harper and Row. German original, 1927.

Husserl, Edmund. 1954. *The Crisis of European Sciences and Transcendental Phenomenology.* Translated by David Carr. Evanston, Ill.: Northwestern University Press. German original, ed. Walter Biemel, 1954 (posthumous).

1989. *Ideas Pertaining to a Pure Phenomenology and Phenomenological Philosophy, Second Book: Studies in the Phenomenology of Constitution.* Translated by Richard Rojcewicz and André Schuwer. Dordrecht: Kluwer Academic Publishers. German original, 1952, from 1915 manuscript.

Melden, A. I. 1961. *Free Action.* London: Routledge and Kegan Paul.

Merleau-Ponty, Maurice. 1962. *Phenomenology of Perception.* Translated by Colin Smith. London: Routledge and Kegan Paul. French original, 1943.

Penfield, Wilder. 1975. *The Mystery of the Mind.* Princeton: Princeton University Press.

Russow, Lily-Marlene. 1988. "Merleau-Ponty and the Myth of Bodily Intentionality." *Noûs* 22(1): 35–47.

Sajama, Seppo, and Matti Kamppinen. 1987. *A Historical Introduction to Phenomenology.* New York: Croom Helm.

Sartre, Jean-Paul. 1966. *Being and Nothingness.* Translated by Hazel Barnes. New York: Washington Square Press. French original, 1943.

Searle, John. 1983. *Intentionality.* Cambridge: Cambridge University Press.

Smith, David Woodruff. 1986. "The Structure of (Self-)Consciousness." *Topoi* 5(2): 149–56.

1989. *The Circle of Acquaintance.* Dordrecht: Kluwer Academic Publishers.

Smith, David Woodruff, and Ronald McIntyre. 1982. *Husserl and Intentionality.* Dordrecht: D. Reidel.

Wittgenstein, Ludwig. 1953. *Philosophical Investigations.* Translated by G. E. M. Anscombe. New York: Macmillan.

5

Background Ideas

Abstract. Here we study the *background* of intentionality and what I call *background ideas.* This chapter shows how everyday experiences depend for their intentional force on ideas extant in one's background culture. These ideas are abstract meaning entities (as in the classical Husserlian model of intentionality), but they arise only in particular historical cultures (they do not exist in a Fregean or Platonic heaven of ideas). And they could not mean what they do without that cultural background. Accordingly, the content of one's thinking, perceiving, willing, and so forth depends ontologically on one's background culture.

Segue. In "Consciousness in Action" we showed that our experience is not isolated within a solitary subject or disembodied mind: our conscious bodily actions are carried out in a physical and indeed social context, and that is part of their intentional content, their volitional meaning. But there is a further, logical reason why consciousness itself normally implicates the social world in which we live. For, as we see in the present chapter, the contents of our intentional experiences are themselves typically composed of concepts and rules of practice that are drawn from and depend on a rich background of ideas that form part of the cultural context in which we live. Normally, then, consciousness is not only embodied but en-cultured. This aspect of consciousness requires a sensitive basic ontology, as we shall explore in later chapters.

The world is so you have something to stand on.
Krauss and Sendak, *A Hole Is to Dig*

For various discussions and/or comments on an early draft of this essay, I thank Christian Beyer, Chuck Dement, Dagfinn Føllesdal, Peter Simons, Barry Smith, Amie Thomasson, and Richard Tieszen. For their contributions to the argument for the background, I am indebted to discussions with Hubert Dreyfus and John Searle. For the opportunity to discuss relevant texts at length, I thank the students in my seminar at the University of California, Irvine, winter quarter, 1994.

Here I want to develop a conception of the *background* of intentionality
and explore the status of what I call *background ideas*. I draw on writings of
Searle and Wittgenstein as well as Husserl, but the framework to unfold –
part phenomenology, part ontology – differs from theirs.

Background

The theory of *ideas* – contents of mental acts or states – weaves through
nearly all of philosophy since its inception. Different strands of theory
come together finally in Husserl's analysis of intentionality, where we find
the most effective theory of ideas yet developed. Drawing on Brentano's
revival of the medieval conception of *intentio*, and Bolzano's distinction
between objective and subjective ideas, Husserl wove together two long
lines of theory: the theory of content in intentional acts of thinking,
perceiving, and the like, and the theory of content in logical assertion and
argument (both types of theory tracing back to Aristotle). On Husserl's
theory, an idea is an objective, abstract ("ideal") entity: not a universal (a
property or type) but a concept, proposition, image – a meaning (*Sinn*).
Its role in thought and/or language is to prescribe (semantically) an
object to which an act of thought or speech refers if this idea is its content.
This doctrine is foundational to Husserl's phenomenology, and it leads
into the role of ideas in consciousness, knowledge, action, language, logic,
and (arguably in today's computer languages) computation.

   Yet ideas are notoriously difficult to get our hands on. Not only are
they abstract entities, nonspatiotemporal and nonperceivable, but they
are normally quite distant from our awareness: only in theoretical reflection do we focus on ideas themselves, in phenomenology or semantics or
psychology. What's more, many of our ideas, about friends or family, are
never brought to consciousness, as Freudian psychoanalysis has argued.
Similarly, Nietzsche's moral psychology unearthed cultural ideas of *ressen-
timent* and "nobility" with an ancient genealogy. Indeed, ideas of value
and culture shape our experience and behavior without our acknowl-
edgment, in ways analyzed by Levi-Strauss, Foucault, Derrida, Kristeva.
More basically, our phenomenological sense of "being" and of "being-
with-others" lies at the periphery of our experience, in ways analyzed by
Heidegger and Sartre. Indeed, our fundamental ideas about the world
are seldom voiced or articulated in any way.

   One's thought or action is directed toward something in the world, as
Husserl said, by virtue of a particular content or idea in one's mind. But
its intentionality is dependent on other ideas in the background of the

act. These background ideas are at work around us all the time, yet we are not aware of them without considerable effort. Such effort we find in the theory of background and of background ideas, which exercise us in the explorations to follow.

The Emerging Concept of Background

A mental act of perception, thought, volition, imagination, and the like is said to be *intentional* insofar as it is "of" or "about" something, and in that sense it "intends" or is directed toward something. On a broadly Husserlian theory, each act of consciousness is performed or experienced *by* a subject and directed *via* a content or idea *toward* some object (typically outside consciousness). The act's content Husserl called a "noema," a technical term (found in Aristotle) for what in modern English we call "idea."[1]

Now, an act can represent or intend something only against a *background* of other phenomena including beliefs, attitudes, norms, and bodily skills. Convergent accounts of background have emerged, in very different idioms, in writings of Husserl, Heidegger, Merleau-Ponty, Wittgenstein, and most recently John Searle and Hubert Dreyfus.[2] I find these accounts convincing but in need of a wider framework of ontology *cum* phenomenology.

I shall lay out a theory of the *background* holding in brief that:

1. Our intentional acts presuppose a *background* of beliefs, skills, emotions, values, social practices, physical conditions including neural states, and more.
2. A crucial part of this background is our fundamental, open-ended *background image* of the world around us, indicating how things are as well as how we do things, even how we use our bodies.
3. This background image consists of *intentional contents*, or *ideas*, including concepts, propositions, values, rules of practice, and items of know-how presupposed in the activities of people in the relevant community.
4. This relation of presupposition between an intentional act and its background is a relation of *ontological dependence*.

The background is, then, part of the *ground* or *substrate* of intentionality: what makes it possible and enables it to work as it does, the preconditions or conditions of the possibility of intentionality. In terms of the notion of dependence or, in Husserl's formal ontology, "founding"

(*Fundierung*): intentionality *depends* or is *founded* on certain background conditions, that is, the act could not be directed as it is – this intention could not exist – unless these conditions existed.[3]

For Searle the background consists of practical capacities or know-how, which takes over where intentional states like belief leave off. What Searle has christened "Background" (with a capital 'B') – namely, know-how – will emerge as only a part of a wider background of intentionality (we will drop the capital B).

Searle's inspiration was Wittgenstein's *On Certainty*.[4] The crucial notion there was that of "fundamental empirical propositions" in our background "world picture." Such "ground propositions" (for instance, that here is a hand) form the foundation of our beliefs about the world around us and merge with our rules of practice. To develop a proper conception of the background, we must see that these ground propositions not only are part of the *epistemological* foundation for various beliefs (conferring evidence) but, along with cognate rules of practice, are part of the *ontological* foundation of various forms of intentionality. A related notion (presumably unknown to Wittgenstein) was Heidegger's notion of "basic" or "ground" concepts, concepts of the ontological ground of human activity (our "being").[5] And the notion of ontological foundation or ground was developed by Husserl, we noted, as part of his formal ontology (and was known to Heidegger).

Wittgenstein's notion of our background world picture can be developed further in terms of Husserl's prior notion of the "life world" or "human world"[6] (a precursor to Heidegger's notion of "world" as in human "being-in-the-world"). Our everyday "sense" of the world forms our background world picture, which depicts the world as we know it in everyday life, the life world. In Husserlian terms, this world picture is a construct of intentional contents or "noemata," embracing concepts, images, propositions, and volitions, modified by attitudes ("thetic characters") of perception, belief, emotion, and the like.

Our background world picture is empirical and highly contingent. It is in many ways a product of human activity, a cultural artifact developed tacitly over thousands of generations. Accordingly, the ontological status of our world picture, of its component ideas or images, is not that of Platonic or Fregean "thoughts" *cum* norms in a Platonic heaven. Rather, their status is akin to artworks as characterized by Roman Ingarden drawing on Husserl's account of cultural objects: they are objectively existing abstract entities (so far Platonic), yet they are brought into existence and maintained in existence by acts of consciousness (no longer Platonic).[7]

In the *background* of our everyday perceptions, thoughts, and actions, then, are ideas wrought by prior human experience and practice, ultimately by prior intentional acts. Intentionality typically thus depends not only on the subject's states of belief, skill, and the like, realized in his or her brain, but also on *ideas* that are extant in the subject's community. These *background ideas* have a life of their own outside any individual's mind: they are (a) abstract meaning entities and (b) cultural artifacts. But if background ideas are abstract, cultural artifacts, ultimately so are all intentional contents or ideas. This crucial point about the ontology of ideas flows naturally, but surprisingly, from the emerging theory of the background.

The Content Theory plus the Background

Let us recount the classical content theory of intentionality and see how it is extended with the notion of background.

The content theory has its locus classicus in Husserl; it has been reconstructed in California phenomenology, and reincarnated in, who?, Searle.[8] According to this theory, an intentional state or act-of-consciousness is experienced by a subject ("I"), has a content, and through its content "intends" or is intentionally related to an object (if such exists). So the structure of an intentional relation between act-subject and object is this:

$$I — act — content \longrightarrow object.$$

The content is a percept, concept, thought-proposition, volition: an *idea*. The content prescribes the object as having certain properties. (The short arrow depicts this part of the intentional relation; the long arrow depicts the whole intentional relation of act by subject to object through content.) In Husserl's version of the story, content mediates the intentional relation much as sense mediates reference in language. In Searle's variation, the content determines the "conditions of satisfaction" of the act-state, the conditions under which the intentional state would be satisfied. (Let us bracket worries about the self and keep a role for ego or "I" in intentionality.)

But the theory of the background says the story is not finished. For an intentional state or act does not occur in isolation but is connected, as Husserl and Searle argue, with a system of beliefs the subject holds about the object or kind of object intended. These beliefs Searle calls the "network," and they are part of what Husserl called the "horizon" of the act.[9] Now, the act *presupposes*, and so its intentionality *depends on*, a network of background beliefs whose contents are associated with the content of the act. In a picture:

network of belief $\vdash$—— < I — act — content ——> object >.[10]

(Here the T-bar depicts the relation of ontological dependence.) The intentional relationship itself depends on a network of beliefs. For instance, when I see yonder bird, my visual experience, whose content is "that gliding turkey vulture," presupposes my background beliefs about birds, for instance, that birds have wings.

But still the story is not finished, says the background theorist. The intentionality of an act depends on a *background* of further conditions that are not tacit beliefs or any other kind of intentional state, but rather...what? Searle proposes a background of *skills*: bodily habits, or dispositions of the nervous system, glossed as "capacities." When I see a turkey vulture overhead (not Searle's example), my visual experience rests on my capacity or ability to turn my head and focus my eyes, my ability to recognize a turkey vulture in flight, my bird-watching know-how. These mind-brain states are not themselves intentional states, but they are an indispensable background of intentionality – in ways to be specified.

And still the story is unfinished. As I watch the turkey vulture overhead, my visual experience presupposes not only various of my personal beliefs and bodily skills, but also a broad complex of communal assumptions about the world in which we live. Wittgenstein called this complex our "background" "world picture." Our world picture, he said, includes our "fundamental empirical propositions," for instance, that here is a hand, that I have two feet, that the earth has existed for a long time.[11] These are "ground propositions" about the world in which we live, propositions that describe the world as we deal with it in basic ways in everyday life. Such *propositions*, Wittgenstein observed, shade off into *rules of practice*, governing not only linguistic activity (language games) but also bodily activity (walking across the room, with my two feet, or sitting in the chair, without checking to see that it has legs). For Wittgenstein, then, the background would consist of fundamental propositions shading into rules of practice. What Searle has done, in effect, is to bring the background rules indoors, from the surrounding community into the individual's nervous system: not as rules followed in unconscious mental execution but as neural states that produce behavior in accord with the rules in the community (rule-governed, not rule-following behavior, as Searle rightly stresses).[12]

Precisely what goes into the background – beliefs, skills, propositions, or norms? When we appreciate the ontology of the background, we should see that this is not an either-or question. Intentional activities may depend, ontologically, on all these types of entities.

When the background is brought in, I propose, the structure of intentionality looks like this:

background ⊢—— < I — act — content ——> object >.

On this *content-background* theory, the intentionality of an act – the relationship binding subject, act, content, and object – depends on a background that includes a network of associated beliefs *and* a system of practical capacities realized in neural pathways[13] *and* a framework of background ideas that form our basic image of the world. In this way the background is part of the operative context of the act, but it is not part of the act's content.

The Argument for the Background

To begin to motivate the notion of background, Searle uses an observation of Wittgenstein.[14] Consider a drawing of a man walking on a hillside. Nothing in the drawing itself determines that it is a picture of a man walking forward uphill rather than a man sliding backward downhill. Only relative to an interpretation, an assumption about what the man is doing, can we see the picture one way rather than the other. Now, the background is like that, Searle suggests: an intentional state can represent one state of affairs rather than another – it can have one set of conditions of satisfaction rather than another – only against a background of "assumptions" or capacities that enable it to represent as it does. The scare quotes warn that, for Searle and Wittgenstein, what seem to be assumptions are often not belief states at all, even deeply unconscious ones, but something akin to "know-how." These issues emerge more clearly as we proceed.

The most compelling argument for the background goes by example.[15] Searle aptly draws this form of argument out of Wittgenstein's *On Certainty*, even though Wittgenstein was arguing in a different direction and had no explicit conception of intentionality. Searle also adapts Wittgenstein's argument about rule following: one cannot follow a rule to follow a rule, on pain of infinite regress. The rule argument, however, applies only to intentional action (which might be mistaken to consist always in following a rule), surely not to perception, emotion, contemplation, or other kinds of intentional state. So we shall develop here the argument by example. This style of argument incorporates a phenomenological minimalism, which argues against overpopulating our phenomenology with too many representations – a stance Hubert Dreyfus derives from Heideggerian phenomenology.[16] To feel the force of this argumentation, we need to describe, patiently, a number of different types of intentional activity.

Consider a simple intentional action like walking:[17]

I am walking, with the conscious experience of so walking.

My action depends on my fundamental knowledge about walking, for instance:

I "know" that the earth does not move as I walk on it.

Is that item of "knowledge" a belief I hold? Is the proposition "the earth does not move as I walk on it" any part of the content of my conscious experience of walking?[18] No. But is it a *belief* I hold, not a conscious thought or judgment but a tacit and unconscious belief, a "silent" background belief presupposed by my experience of walking, including my intention or volition in walking? Perhaps, but consider what happens next:

As I am walking, an earthquake strikes the area and I feel the earth rolling, like a wave beneath me, as I try to walk on.

Before the earthquake, before my rock-and-roll experience, I did not have an *active* belief to the effect that the earth does not move. Of course, I knew about earthquakes and I had an intellectual belief something like that. But that belief played no role in my experience of walking. It was not even a background belief *presupposed* by my experience of walking. This is debatable, but as we expand the range of ostensible background beliefs, it grows gradually implausible that all this cognitive structure is really used somehow, genuinely presupposed, by my simply walking along: on the ground, on the surface of the earth – where else?

Maybe this sort of fundamental knowledge or "belief" is a skill, an item of know-how rather than knowing-that. That is the line taken by Searle and by Dreyfus (Searle drawing on Wittgenstein and Dreyfus on Heidegger and Merleau-Ponty). Still, I think it is not right. To see why not, we have to dig into Wittgenstein's original intuitions about the background. This sort of "belief," we shall find, is somehow deeper than the beliefs carried "in my head": it is part of our communal world picture, on which my belief states themselves rest.

Take another example, not an action but a simple visual experience:

I see that turkey vulture gliding over the meadow.

Now:

I believe that birds fly through the air.
I believe that wings enable birds to fly above the surface of the earth.
I believe that the earth has an atmosphere of nitrogen, oxygen, and
so on.
I believe that the earth is a globe floating in space.
I believe that the earth is very old.

Are these beliefs part of the network of background beliefs presupposed by the content of my visual perception? Perhaps the first two, but not the last ones. Although I do believe these things, they play no role in my visual experience. As the content of my experience fans out to associated content, to background presuppositions, the horizon of my perception soon reaches a boundary beyond which the "propositions" that seem required in order to make sense of "what I see" are either abstract or scientifically informed beliefs or, well, not really beliefs at all. My wider knowledge of the world takes over, some merely practical and some pointing to the knowledge of experts. I know that I cannot fly, that birds can. But this "knowledge" is as much know-how as know-that, as much practical skill as theoretical belief. The problem of the background (in Searle's sense, with an eye to Wittgenstein) begins with the problem of where to draw this boundary, where bona fide intentional content leaves off and practical know-how or skill takes over.

Consider now a purely intellectual, conscious activity of thought. Reflecting on a recent earthquake,

I think that the earth does not normally roll beneath me as I walk.

There are a great many collateral beliefs I would seem to hold:

The earth is a mass of dirt I may walk on.
The earth has hills and valleys I may walk on.
The earth is molten iron covered by a crust of dirt.
The earth is a roughly spherical body.
The earth is one of nine planets that orbit the sun.
The dirt in California is a dry powder in summer.
The earth has existed for a long, long time.
There has been no rain hereabouts (in California) for some time.

There is no mud underfoot as I walk across this hill.
I weigh 175 pounds.
I weigh the same as my brother.
I move my feet in this way [an image] when I walk uphill.
I have two feet.
I use both my feet when I walk.

I believe all of these things. Yet which of these propositions are contents of bona fide beliefs in the background of my consciously thinking "the earth does not normally roll beneath me as I walk"? Is it that I actually believe all of these things, or only that I would reach such conclusions quickly if I thought about such things? If these really are actual beliefs I hold, albeit far from my consciousness in thinking "the earth does not normally roll...," do they play any role in my so thinking? Are these beliefs engaged as background beliefs of my so thinking? Are they presupposed by my act of thinking, so that my thought actually depends on them?

It is implausible that my thought actually engages – presupposes and so depends on – all these beliefs. Potentially perhaps, but not actually. The point is that sooner or later, as we list propositions to which I would assent or propositions to which I could be expected to reason rather naturally were I to think on and on about the earth and my walking – sooner or later, it becomes implausible that all these beliefs are a proper part of the system of background beliefs on which my thought actually depends. And yet, we want to say, they are there, implicitly at least.

The preceding line of argument for the background has a form we may summarize as follows. First, assume that an intentional act has a content (which prescribes the object of consciousness). Second, from examples observe that an act presupposes a variety of collateral beliefs or knowledge as well as cognate practices and know-how. Third, from plausibility considerations observe that these items are too much or too inappropriate to be in the content itself. Finally, conclude that they form a background presupposed by the act but not explicitly included in its content.

The background of intentionality, then, includes our fundamental "knowledge" or "assumptions" or "attitudes" about the world with which we are dealing. But what sort of things are these "assumptions"? It remains to specify more carefully the ontology of the background, the type of entities on which intentionality depends in the ways described: the background ideas we regularly presuppose.

Background Capacities or Know-How

The "assumptions" that form the background of our everyday intentional activities outrun the range of *beliefs* plausibly presupposed by our activities. Searle defines the background, in cases similar to those we considered, as composed of "capacities." In *Intentionality* Searle wrote (1983: 143):

The Background is a set of nonrepresentational mental capacities that enable all representing to take place. Intentional states only have the conditions of satisfaction that they do, and thus only are the states that they are, against a Background of abilities that are not themselves Intentional states. In order that I can now have the Intentional states that I do I must have certain kinds of know-how: I must know how things are and I must know how to do things, but the kinds of "know-how" in question are not, in these cases, forms of "knowing that."

The term Searle prefers for elements of the background is "capacities" or "practices" (p. 156). Terms like "assumptions" or "presuppositions" are misleading because they make background capacities sound like *representational* states, whereas these capacities are rather a kind of "know-how." Yet these capacities, Searle insists, are "explicitly mental."

In *The Rediscovery of the Mind* (1992) Searle modified his conception of the background. Consciousness, he there argues, is the defining character of mind: though caused by and realized in the nervous system, it is an irreducibly "subjective" property of the mental. Accordingly, the background consists of dispositions of the nervous system: states of know-how are unconscious capacities and so are properly identified with neural dispositions and only misleadingly called "mental." And the network of beliefs Searle now places within the background: what we have wanted to call unconscious beliefs (from psychoanalysis to cognitive science) are better cast as dispositions of the nervous system – dispositions to produce conscious intentional states of thought or judgment. (Ironically, Searle notes, this turns Ryle upside down, or rather inside out, or outside in: "mental" states are dispositions to produce not behavior but consciousness.)

Clearly, however, the mental background of intentional states is wider than what are appropriately called skills or capacities. Searle takes a step toward widening the background when, considering collective intentions, he writes:[19]

In addition to the biological capacity to recognize other people as importantly like us, in a way that waterfalls, trees, and stones are not like us, it seems to me that the capacity to engage in collective behavior requires something like a preintentional sense of "the other" as an actual or potential agent like oneself in cooperative activities.

Now, our "sense" of things – of ourselves, of others, of our times and our life world – lies partly in our preintentional practices or ways of dealing with things. But it also lies partly in our intentional attitudes toward things, in our *image* of our world, which is not reducible to skills or know-how. Accordingly, we need a wider notion of background than what Searle dubs "Background," or even background of skills *cum* network of beliefs. I want to reclaim the term "background" for this wider range of enabling conditions of intentionality.

Our Background World Picture

Wittgenstein's *On Certainty* Searle declares "one of the best books" on the background.[20] Indeed it is, and it leads to a trenchant notion of background. What I find most illuminating, along with the many concrete examples, is Wittgenstein's conception of our background "world picture." For only against our basic if inchoate image of things can our intentional states represent as they do. To those who read Wittgenstein as deeply antimentalist, however, as squeezing the intentionality out of language and even thought, I shall be reading Wittgenstein against himself as I extract his account of our world picture as background of intentionality.

In *On Certainty* (remarks dating from 1949 into 1951) Wittgenstein considered the kind of empirical propositions (*Erfahrungssätze*) that Moore said we can know with certainty, for example, that "here is a hand." These fundamental empirical propositions, Wittgenstein observed, form our background "world picture" (*Weltbild*).[21] They are the "ground" (*Grund*) of our language game of questioning or doubting, against which our whole practice of so doubting makes sense; in this role they are immune from doubt. But the term "ground" has two meanings: our grounds are our reasons or evidence for empirical beliefs, but they are also the "foundation" (*Grundlage*) or "fundament" (*Fundament*) of our empirical beliefs. In Husserlian terms, they are "founded" or depend on, and in that sense are presupposed by, our intentional activities of dealing with the world around us, for instance, in walking, in seeing a gliding turkey vulture, or in consciously thinking or judging about a bird maneuvering in flight.

Wittgenstein says that our fundamental empirical propositions merge with *rules* of their language game, part of their form of life. That is, bona fide propositions shade into a background of practical rules or norms. For Quine, there is no sharp boundary between analytic and synthetic propositions, between logical and empirical propositions. Similarly, according to Wittgenstein, there is no sharp boundary between fundamental empirical propositions and rules – rules of practice concerning our speaking about and our dealing with things described by such propositions. (Remember that Wittgenstein took mathematical and logical propositions as conventions: not claims of truth but decisions on how to talk.) So, for Wittgenstein, many of the fundamental empirical propositions that articulate our world picture are not really *propositions*. That is, what plays the role of grounding our thoughts and actions regarding the world are not propositions that are contents of our beliefs but rather something *outside us*, namely rules of our language games, rules of practice in our community. We learn, usually not explicitly, to think and act in accord with those rules, with the forms of life in our community. Only in that sense do we all *know* that we have hands or that the earth has been around a long time. To simplify, this knowledge begins in propositional knowledge but blends into know-how.

To carry such a line of argument further, in a different vein than Wittgenstein himself did: it is not that we could not explicitly think or judge such things – we could. But those thoughts do not serve to ground our actions. They are the philosopher's retrospective reflections. Even where we hold such bona fide beliefs, because we are either philosophers or articulate artisans, these beliefs are not *engaged* in our everyday activities of moving about the earth. There is no need for our holding such thoughts, even in the form of unconscious beliefs, in order to act as we do. We simply evolved in accord with a world of which these propositions are true. We take these things for granted, but many of them we have never thought about, nor have we need to. Thus our thought and speech rest on a background that runs from fundmental empirical propositions that "everyone knows" to propositions accessible only through belabored reflection, and on to rules of language, thought, and action that no one need ever formulate explicitly.

There is an important difference, however, between Wittgenstein and Searle on the background. For Wittgenstein, the elements of the background that outrun bona fide propositions are rules of practice, norms extant in the individual's community. For Searle, however, the background that outruns beliefs consists of practical capacities realized in one's nervous system – if we may put it so, as dispositions to act and indeed to experience in ways governed by the community's rules of practice. Of course, the bona fide propositions that Wittgenstein finds in the background are intentional contents of the belief states that Searle initially called the "network" and later subsumed under the "background." However, the rules of practice Wittgenstein cites in the background cannot, for Searle, be intentional contents of our background capacities, because exercising these capacities cannot consist in following rules, on pain of infinite regress (Searle following Wittgenstein on rule following). Accordingly, Searle repeatedly stresses, the background capacities that enable intentional states to represent are themselves nonintentional, or preintentional. Thus, both background propositions and background rules of practice lie outside the system of concrete mental-neural states that Searle defines as "background." Yet they are surely parts of the enabling background of intentionality.

Wittgenstein's notion of "form of life" (*Lebensform*) invites a still wider conception of the background. The fundamental empirical propositions that form our world picture might be cast as elements of communal knowledge. Think of such propositions not in a Platonic-Fregean way as eternal objects that represent states of affairs in the world. Think of them rather as elements in a communal database, elements of knowledge formed by one's community as a whole. Sometimes they reside in explicit beliefs in the minds of the experts, as in modern science or medieval theology or Roman rhetoric. But often they reside rather in the practices of the community, available for explication by philosophers in due time but normally only implicit in our forms of intentional activity, from bodily and social behavior to speech to thought. The background of our intentional activities, then, is our form (or forms) of life, embracing our fundamental empirical propositions *cum* rules of practice. Of course, this is something quite different in ontology than the neural dispositions that realize our capacities to behave and indeed to think in conformity with our form of life.

There are intimations of the background already in Wittgenstein's first work, the *Tractatus Logico-Philosophicus* (1921). (*On Certainty* was written during his final year and a half.) In the Tractarian model, the world is built up from facts, or existing states of affairs, which we picture in propositions, and a proposition represents a fact only if they both have the same "logical form." In Wittgenstein's later view, "form" is transplanted from *logical form* to *form-of-life* – cultural form, as language is taken as part of a wider range of human activity. Now, whether form is a feature of the world to which language and other activities conform, or whether our forms of life, language, and even "world" are human artifacts, form is shown but not said. Form is thus presupposed but not represented in thought, language, and action. In our terms here, forms of language or life are part of the background of intentionality.[22]

So far we have unearthed very different kinds of entities in the background of intentionality: *propositions* in our world picture, *rules* of practice that define *forms* of life and language in our community, and our individual practical *capacities* (the latter inscribed in our nervous system). There are close connections among these things. All are involved in our basic image of the world around us, an image that is largely communal, the product of eons of human experience (and in some places informed, increasingly, by the knowledge of experts such as scientists). And that image, I submit, forms a crucial sort of background for our intentional activities, without which our thoughts and experiences could not have the intentional force they do.

But what sort of thing is our *image* of the world? Here I want to mark out the background *ideas* that are so important to our daily lives.

Our Background Sense of the Life World

The notion of our background world picture grows clearer as we look at Husserl's account of what he called the "life world" (*Lebenswelt*) or the "human world."[23]

In the *Crisis* (1935–38) Husserl says the life world is the "ground" of all human activities, theoretical or practical, whence the ground of all (human) intentionality. What is the life world? It is the world as experienced – and *assumed*, merely taken for granted – in everyday life, in everyday intentional activities, from perception, thought, imagination, and emotion to action. But what is the world-*as*-experienced? There are two candidates: the world in which we live, our surrounding world itself, restricted to certain properties; and our conception or image of the world, our "world picture" (*Weltbild*), as Husserl calls it. The world itself is external to consciousness, whereas our image of the world is a construct of intentional content, what Husserl called "sense" (*Sinn*) or "noema." Our everyday, commonsense, life-world image of the world is the image that informs our everyday activities. This is to be distinguished from our scientific image of the world, though our everyday life-world image both shapes and is shaped by our scientific image. (Sellars's influential distinction between the "manifest image" and the "scientific image"[24] recapitulates Husserl's distinction between our experiential, life world conception and our theoretical, natural-science conception of the world.)

Which then, according to Husserl, is the *ground* of our intentional activities: the life world itself or our sense (conception) of the life world? Husserl's story is complex.

The principles of Husserl's account of the life world were laid out in his *Ideas* II (1912). In *Ideas* I, laying out his phenomenology and associate ontology, Husserl distinguishes "sense" (*Sinn*) from "essence" (*Wesen*): the essence of something consists of the properties that make it what it is, whereas the sense of an intentional act is the content that represents or prescribes its object as having various properties. In *Ideas* II Husserl describes one world of objects, which may have different aspects or essences, that we experience or intend through different conceptions or ranges of sense. Specifically, he analyzes our conceptions of the natural, cultural, and subjective-intentional essences of things in the world. These very different ranges of essence Husserl calls Nature, Humanity or Spirit (*Geist* in the sense of *Zeitgeist*), and Consciousness.[25] Strictly speaking, then, the Life World is neither the world in itself nor our everyday conception (sense) of the world, but that aspect (essence) of things in the world formed by those of their properties that we deal with in everyday life: the properties that define the Human or "Spiritual" World. For instance, this material object before me (a pen) is composed of quarks in a complex configuration: a naturalistic property. It also is used for writing: a cultural property. It is not conscious, so it has no subjective properties, no intentional experiences (such as thinking of what I am writing with it).

For Husserl, then, our *sense* of the life world prescribes the *essence* of the life world, that is, the Life-World essence of things around us. According to Husserl's detailed analysis, the essences of things are stratified in the following way,[26] as things in the world around us may be respectively:

   (i) spatiotemporal objects,
   (ii) material objects in causal relations,
   (iii) living things, including animals and humans with psychological states,
   (iv) human beings proper, or persons, bearing culture and ethics,
   (v) cultural objects such as houses, books, laws, and institutions produced by human beings, or
   (vi) subjects of intentional activity or "pure" consciousness (as distinct from natural, psychological, or cultural phenomena).

The essence Nature comprises levels (i)–(iii); the essence Life World or Humanity, levels (iv)–(v); and the essence Consciousness, level (vi).

According to Husserl, these levels of essence are prescribed by corresponding levels of sense, through which things are presented with properties belonging to these levels of essence. Thus, our everyday *sense* of the world presents things with Life-World *essences* such as being a person or a chair, while propositions in natural science present things with essences of Nature, and propositions in phenomenology present mental acts as having intentional structures or essences of Consciousness.

On Husserl's account, there are relations of *dependence* among these levels of essences – or, alternatively, among the instances of these essences in particular objects. Husserl finds the later entries dependent on the prior ones, and recent accounts of the "supervenience" of mental events on physical events would concur, provided supervenience entails dependence.[27] The details of these dependencies, however, are highly controversial. In his turn to transcendental idealism (*Ideas* I, §49), Husserl claimed there is another dependence that runs in the other direction, in the way consciousness gives meaning to what is intended, say, as a lump of clay, an animal, a human being, or a book. In that way properties at levels (i)–(v) depend on intentional properties at level (vi). This form of transcendental idealism, or internal realism, is not our concern here. Our concern is rather the ways in which intentionalities, at level (vi), depend on various levels of properties of things, from (i) through (vi). Thus, I cannot perceive a chair (have an experience with the intentional property of seeing-a-chair) unless I have dealt with chairs (things with the Life-World property of being a chair) and indeed acquired from my culture the sense "chair," which prescribes the property or essence of being a chair. In both these ways, then, intentionality is dependent inter alia on the life world.

Now, our *sense of the life world* – our commonsense image of the world as we encounter it in everyday life – is precisely what we have here called our *background world picture*, the background of our familiar forms of intentionality.[28] Husserl's detailed analysis thereof brings out crucial elements of our awareness of ourselves and others. Our bodily awareness in kinesthesis and motor control or will come with our awareness of ourselves as animated beings along with other animals. Empathy with others and thus social awareness enter with our awareness of animal life and are crucial on the level of human beings or persons. And our sense of values comes with our awareness of our selves and other persons living in a culture.

In Husserlian terms, our background world picture would be a system of *background ideas* – *senses* or *noemata* – that prescribe features of things in the world as we experience them in everyday life. This picture would embrace all the features in (i)–(vi), and properties involved therein such as kinesthesis, motor control, empathy, value, and more. Taking their places in this picture are inter alia the "background" items stressed by Searle and Wittgenstein: our practical capacities, including bodily, social, and intellectual skills; the rules of practice in our forms of life, extant in our culture; and our fundamental propositions about various kinds of things, so far as these are either commonly voiced or reasonably accessible on reflection. Our world picture should also include our fundamental values, so far as these are involved in both our forms of life and our assumed or voiced "beliefs" about what is important, obligatory, and so forth.

Husserl sometimes characterized the surrounding world as the "horizon" of an intentional experience. To speak precisely, the *horizon* of my seeing yonder house is the range of possibilities left open by the content of my perception.[29] These are possibilities for the world around me: starting with properties of the back side of the house, also the inside, and ranging from physical to cultural and aesthetic properties, thus from the Natural World to the Life World. These possibilities, Husserl said, must be "motivated," not "empty," possibilities. In present terms that means they must be compatible with our background world picture. But while the horizon of an act goes on indefinitely, there are limits reflecting our background world picture, which is open-ended and revisable, yet limited. Where Wittgenstein said our background knowledge runs out and assumed propositions blend into rules of practice, Husserl said there are potential lines of further experience, which prescribe the horizon by specifying how knowledge can be extended through appropriate action (walk around the side of the house and you will see if there is another porch). Consequently, it belongs to the horizon of our experience that we can develop new forms of action, practice, language, knowledge, even perception. Accordingly, our basic sense of the world, our world picture, which constrains the horizon of our experience, includes both practical and cognitive intentional contents indicating the revisability of our background picture itself.

The Background as Ontological Foundation

How exactly is the intentionality of an act of consciousness related to the act's background? The act's intentionality, I propose, *depends* or is *founded* in certain ways on various parts of its background. So the relation between intentionality and background is that of ontological *dependence*, and the role of background in intentionality is that of ontological *foundation*.

Drawing on Husserl drawing on Aristotle, we say:

A *depends* or is *founded* on B if and only if A could not exist unless B existed, that is, necessarily A exists only if B exists.[30]

Dependence is not in general "causal." Aristotle held that this whiteness in this paper could not exist apart from the paper, but the paper is not a *cause* of the whiteness in our modern sense of "cause." B is a "causal" condition of A only in case the necessity involved is physical, and a brain event B is a "neural" condition of an intentional act A, *pace* Searle, only in case the physical necessity concerns neural activity in the subject's brain. Where intentional activity depends on background ideas or rules-of-practice, *pace* Wittgenstein or Husserl, the dependence is not causal and surely not neural. The wide notion of ontological dependence, then, allows us to define the dependence of intentionality on different background phenomena, including background ideas and norms, without reducing the dependence to causation.

Assume this traditional analysis of ontological dependence. Then, I propose: the intentionality of an act of consciousness *depends* or is *founded* on its background, in that the intentional relationship among subject, act, content, and object could not exist – the act could not be directed as it is – unless the background conditions or entities existed.[31]

We may schematize the ontology of intentionality, then, as shown in Figure 5.1. The point of this figure is to stress the different roles of background and content in the intentional relationship of act to object: the content *mediates* the intentional relation (by semantically prescribing the object), whereas the background *founds* the relationship (by making it possible).

This dependence on background is not the dependence of a mental state on a brain state, as in recent theories of the "supervenience" of mind on brain.[32] Rather, the dependence at issue is that of a mental act's *intentionality* on its *background*: the act could not be *directed* as it is unless the relevant background phenomena existed. Moreover, the ontological

```
                    experiences      invokes           prescribes
background |— < subject ———————— act ——————— content ——————— > object >
      founds                      intentional relation
```

FIGURE 5.1. The ontology of intentionality.

foundation of intentionality includes not only the neural activity in the
subject's brain but also his background ideas about the world, ideas extant
in his culture, not in his brain.

Where the background entities are ideas or propositions in our world
picture, we cannot follow Searle's characterization of the background as
"causal" conditions. It is useful to recall Searle's characterization, though,
to see how it fits into our wider, more robust conception of background.
In *Intentionality* Searle characterized the role of background, restricted
to practical skills, as that of "enabling conditions" (1983: 158):

[T]he Background [is not some further Intentional content that latches on to
the primary Intentional content, but] is rather the set of practices, skills, habits,
and stances that enable Intentional contents to work in the various ways that they
do, and it is in that sense that the Background functions causally by providing a
set of enabling conditions for the operation of Intentional states.

In *The Rediscovery of the Mind* Searle said, more simply, that "Background
capacities that are not themselves intentional" (1992: 175) but "enable
our [intentional] mental states to function," that is, to represent (p. 175):
they are, in a Kantian idiom, "conditions of the possibility" of intentionality (p. 195). In *The Construction of Social Reality* Searle says these
Background *capacities* are "a certain category of neurophysiological causation" (1995: 129). Accordingly, "*Enabling* is meant . . . to be a causal
notion. We are not talking about logical conditions of possibility but about
neurophysiological structures that function causally in the production of
certain sorts of intentional phenomena" (p. 130). So Searle conceives
background capacities as neural, not logical, conditions of possibility,
and he takes these as a kind of *causal* enabling condition. We, however,
must define "enabling" in a way that does not reduce to causation but allows causal conditions as a special case of enabling conditions. So what is
meant by an "enabling condition" or "condition of possibility"? The only
clear accounting I know of uses the notion of ontological dependence,
as explained earlier.

The ontology of background clarifies issues of externalism in recent
philosophy of mind. *Internalism* holds that the intentionality of a mental
state is determined by content "in the head" (= mind); Husserl's theory
of intentionality is internalist and so is Searle's, but both are ramified by
notions of background. *Externalism* holds that intentional content is determined not by what is "in the head" but by external conditions of causal or
social context – by what we have called background conditions; on Tyler
Burge's view intentional content is "individuated" by social context,[33]

whereas on Fred Dretske's view perceptual content resides in the physical "information" in the causal route from object to perception.[34] The ontology of externalism would thus fuse the content and context or background of an intentional act and so conflate the importantly different roles of background and content. Internalism, on the other hand, tends to either omit the background or fill our heads with too many ideas aping the background. Now, the theory of background we have sketched avoids these errors. We put context where it belongs: in the background! – and content where it belongs: in the head![35]

The Ontology of Background Ideas

What is the *ontological category* of intentional contents, or ideas: concepts, thoughts or propositions, images, volitions, and the like? According to a long tradition noted earlier, ideas are not mental events or subjective "appearances," but objective entities extant in the world. They are shareable as contents of different acts by different people, and they are communicable in language, art, and other media. Yet they are not physical objects, and they are not themselves located in time or space. From the Stoics to Bolzano, Husserl, and Frege, ideas have been conceived in this way as objective abstract entities. We need not put them in a Platonic or Fregean heaven outside the "real" world, however. In fact, the whole thrust of Wittgenstein's examples of "fundamental empirical propositions" is that these propositions are contingent upon our earthly human existence. Their presence in the *background* of our experience in this world requires that they not be confined in an ideal heaven. They are abstract – Husserl said "ideal" – entities, yet they are in this world, available to our intentionalities.

It is vital that our intentional activities share much of the same background ideas or "assumptions," both practical and cognitive. Without such shared background, communication is impossible. Science fiction, describing encounters with extraterrestrials, illustrates what we all well know about the background of intentionality; but the lesson is familiar much closer to home. The point to notice here is that the background must include *shareable* objective intentional contents, roughly of the sort posited by Bolzano and Husserl. And these background contents cannot, then, be identified with token states of one's nervous system (as Searle proposed), though they are surely implemented thereby. Nor are these contents one's own personal skills or capacities (as both Dreyfus and Searle might suggest). In the case of practical background, the contents are shareable items of know-how, which we may learn from one another, and which our personal skills instantiate as we acquire the same skill our fellows have. These principles of know-how are rules of practice extant in our culture. They may be operative even if no one ever articulated them explicitly, as in rules of body language such as how to move in walking down the street (say, quick-paced, bouncily, with a swagger, or carefully, soft-footedly, with restraint). Or they may be operative precisely because people articulated them in a law, as in rules of driving such as "Turn right on a red light if no one is coming through the intersection" (a law instituted in California but not in some other states). The important point is that such rules are "out there," extant in our culture.

Now, our background world picture has just this status as an objective artifact of intentional, cultural activity. It consists of ideas – images, concepts, propositions, feelings, values, and items of know-how – that are extant in our culture, realized in our form of life, shared background in our diverse activities. It is a product of human experience, thought, action, and practice, formed over hundreds of thousands of years, refined over a couple of thousand years of "history."

Our scientific theories are increasingly a special part of that picture. But our everyday image of our kind and our world is a still more fundamental part, our life-world image, presupposed yet modified by our scientific image. Indeed, our most basic ideas of things have a status like that of what Carl Jung called "archetypes," primordial images written into our genetic *cum* cultural heritage or "collective unconscious" and articulated in explicit cultural works.[36] The Jungian archetypes, for instance of the masculine and feminine or of the "shadow" side of a person, remind us just how contingent our world picture is. More recently, biologist Richard Dawkins has coined the term "meme" for cultural ideas that evolve and are inherited socially in a pattern similar to the evolution of physical traits. Daniel Dennett, in *Consciousness Explained* (1991), has urged a role for such "memes" in the structure of consciousness, and I would agree. But I would insist on a more robust account of consciousness than Dennett's neo-Rylean "intentional stance" allows, and a more articulate ontology of these entities, our background ideas.

In Sum, Ideas Rule

The world as we know it is built around the ground we stand on, literally, as we walk and talk and live and breathe. That is a fact external to our consciousness in walking, talking, thinking. Nonetheless, an idea of that fact – a "ground" proposition – is part of our background image of the world. Not only does my walking depend ontologically on the ground's stability, but the intentionality of my volition in walking depends onto- logically on the proposition that the "world" affords a stable ground on which we walk. That proposition and related rules of practice belong to the background of ideas or "groundrules" assumed in our culture, indeed, in our species. And that *background* is a crucial part of the ontological *ground* of our activities as members of *Homo sapiens sapiens*.

Notes

1. This theory of intentionality, discussed later, is developed, in reflection on Husserl's work, in Smith and McIntyre 1982.
2. See Searle 1983: chap. 5, 1992: chap. 8, and 1995: chap. 6; Wittgenstein 1972/1949–51; Husserl 1962/1913, 1989/1912, and 1970b/1954. All are discussed here. See Dreyfus 1991, for an interpretation of Heidegger's *Being and Time*, stressing the role of background practices and skills in human activity or "comportment" (*Verhalten*).
3. Husserl's notion of founding (*Fundierung*), or dependence, was laid out in the third of his *Logical Investigations* (1900–1), elaborating a conception orig- inating in Aristotle. A version of this notion of dependence or foundation is applied to background conditions that serve as grounds of intentionality, specifically "direct" awareness, in D. W. Smith 1989: chap. 6.

   The conception of formal ontology, distinguished from material ontology, has roots in Aristotle and Leibniz. However, it seems to have been explicitly developed first by Husserl, beginning in the third of his *Logical Investigations* (1900–1) and given a sharp focus in the opening chapter of his *Ideas* I (1913). Wittgenstein's *Tractactus Logico-Philosophicus* (1921) can be read as detailing a formal ontology: the world is composed of facts, that is, existing states of affairs; these facts have certain *forms* (shared by the propositions that rep- resent them); and the contents of these facts are objects and properties or relations of various kinds that define, in Husserlian idiom, different "mate- rial" domains. Which domains Wittgenstein had in mind (sense data, physical objects, etc.) is a notorious problem in interpreting the *Tractatus* – which just shows that Wittgenstein had his hands on a formal ontology. Indeed, Husserl's formal ontology also recognized, independently of Wittgenstein and earlier, the formal objects called states of affairs (*Sachverhalten*), which he called "cat- egorial" objects "syntactically" formed. A recent study in formal ontology, in Husserl's sense, is Johansson 1989.
4. Wittgenstein 1972/1949–51.
5. I shall not have room to deal here with Heidegger's "ground concepts," which I see as forming a "deep background" of intentionality. See Heidegger 1993/1941. Compare Heidegger 1969/1929 and also 1991/1957. The German word *Grund* is translated sometimes as "ground" and sometimes as "reason." Heidegger himself appraised the different meanings of the word, shifting from the epistemological to the ontological notion of "ground."

6. Husserl's notion of the "life world" (*Lebenswelt*) is central in *The Crisis of European Sciences and Transcendental Phenomenology* (1970b/1954/1935–38). Closer detail is found in his account of the "spiritual world" (*geistigen Welt*), also called the "human" (*menschlich*) or "personalistic" world, in *Ideas* II (1989/1912). Husserl's conception of the human world was a precursor to Heidegger's notion of "world" as in human "being-in-the-world."

7. See Ingarden 1989/1961 and 1964. A recent study of the ontology of fictional entities, revising and improving upon Ingardenian principles, is in Thomasson 1998.

8. See Føllesdal 1982/1969; Smith and McIntyre 1982; Dreyfus 1982; and Searle 1983.

9. Searle originally distinguished the Network of background beliefs from the Background of practical capacities; see *Intentionality* (1983), chap. 5. In *The Rediscovery of the Mind* (1992), chap. 8, he argues that the Network is a special part of the Background, collapsing unconscious beliefs, and hence the Network, into neural dispositions in the Background. In *The Construction of Social Reality* (1995), chapter 6, Searle again defines the Background as preintentional neural capacities. Husserl's notion of horizon is reconstructed in Smith and McIntyre 1982: chap. 5. See chap. 5, sec. 3, on the role of background beliefs, ranging from general beliefs (say, about birds) to fundamental ontological beliefs (say, about material objects) to concrete beliefs about a particular individual. Husserl defined the horizon of an act in different ways, none of which coincides with the definition of either network or background as discussed here but which lead in clear ways into the issues of network and background. Husserl defined the horizon of an act as the range of possibilities left open by the content of the act. These possibilities are empirically "motivated"; they are constrained not only by the explicit content of the act but also by the "implicit" content, that is, the content of relevant background beliefs.

10. This kind of presupposition, taken as precondition or "ground" of intentionality, is analyzed in D. W. Smith 1989: chap. 6. The kindred notion of logical presupposition, a special form of dependence in Husserl's sense, is noted in Simons 1987: 290ff.

11. Wittgenstein 1972/1949–51: 1ff. 94–96, 136, 151ff., 202ff., 308–9, 401, 411. Discussion of the key points follows.

12. See Searle 1995: 139–47.

13. Searle 1983 separates the Network (beliefs) from the Background (skills), where Searle 1992 incorporates the Network into the Background, where both background "beliefs" and background skills are built into the nervous system.

14. See Searle 1992: 177.

15. In *Intentionality* (1983) and again in *The Rediscovery of the Mind* (1992), Searle argues for the Background partly by analogy with issues of linguistic meaning and metaphor. Here I want to stay close to the problem of intentionality itself. Indeed, the most persuasive argument for the hypothesis of the background is a detailed description of simple cases of intentionality, along lines charted by Searle in both books.

16. Compare Dreyfus 1991. I think that Dreyfus is right in cautioning against putting too much intentional content in very basic forms of experience but wrong in two ways. First, he sometimes seems to identify intentionality itself, or the very relation of subject to object, with too much content. Second, he gives too big a role to skills or "coping" because he does not have a variegated notion of background.

17. This is a variation on an example recounted by Searle 1992: 184–85. In Searle's example a famous European philosopher was walking with Searle on the Berkeley campus, discussing the notion of background, when an earthquake struck. The visitor realized after the quake that before then he had not held a belief that the earth does not move; he had simply taken this for granted. This convinced him of the hypothesis of the background.

Assume that walking is an action, a complex event or process that includes as a proper part a volition (an intentional state or event) with a certain content that represents my bodily movement in walking. An account of this structure of action, informed by a content theory of intentionality, is detailed in my "Consciousness in Action" (1992). My account there is consonant in many respects with Searle's account of action in chapter 3 of *Intentionality* (1983); the differences are that I assume a more pointed version of the content theory of intentionality and I detail the phenomenology of the background states of perception and kinesthesis typically associated with the volition (for Searle, intention-in-action) constituent in an action.

18. A conscious experience of action involves a variety of interrelated experiences, including volition, kinesthesis, and perception. These may presuppose beliefs with various contents. Such beliefs, however, are not occurrent events of consciousness but "background" attitudes, which on some analyses are dispositions to have occurrent conscious thoughts. On the phenomenological or intentional structure of action, see D. W. Smith 1992.

19. See Searle 1990: 413.

20. Searle 1992: chap. 8, n. 1.

21. Wittgenstein used the term *Weltbild* as early as in the *Tractatus*, his first work, as well as in *On Certainty*, his last work. He borrowed the term from Heinrich Hertz's *Principles of Mechanics* (1894), which urged that science owes us a "law-conforming world picture" (*gesetzmässig Weltbild*). I am indebted to Joseph Tougas's study of Wittgenstein's use of ideas from Hertz, "Hertz und Wittgenstein" (1996).

22. On the limits of representation, see Hintikka and Hintikka 1986 and Kusch 1989. The Hintikkas argue that, for Wittgenstein (early and late), language cannot contain its own semantics, cannot represent its own form, which is a ground or enabling condition. In the Tractarian scheme, as I should like to put it, form is part of the background, and ground, of representation in language, even if it can be represented in a limited range of language called logic, or in a separate metalanguage. Kusch develops these themes in the systems of Husserl, Heidegger, and Gadamer.

23. See Husserl 1989/1912 and 1970b/1954. See Føllesdal 1990. Much of the detail of the account to follow draws on D. W. Smith 1995.

24. Sellars 1963.

25. See D. W. Smith 1995. These distinctions are detailed, in diffuse discussion, in Husserl 1989/1912. Husserl began using the term "life world" (*Lebenswelt*) around 1917, borrowed from Georg Simmel. In *Ideas* II he called the essence of the life world either Spirit (*Geist*) or Human (*Mensch*).

26. This structure is reconstructed in D. W. Smith 1995, drawing out ideas in Husserl's *Ideas* II (1912).

27. See Kim 1993, especially the conceptual survey of different notions of supervenience in the essay "Supervenience as a Philosophical Concept." Kim discusses different conceptions of supervenience, only some of which involve dependence.

28. On kinesthesis, empathy, and value in Husserl's scheme, see D. W. Smith 1995. A detailed discussion of Husserl's analysis of our commonsense ontology, and also physics, is found in Barry Smith 1995.

29. This conception of horizon is reconstructed, and related to recent possible-worlds theory, in Smith and McIntyre 1982: chap. 5.

30. Husserl's notion of founding (*Fundierung*), or dependence, was used in many later works after being developed in the third of his *Logical Investigations* (1900–1), elaborating a conception originating in Aristotle and used by Brentano. The Husserlian analysis is assayed in Barry Smith 1982. A wider discussion of ontological dependence, amplifying Husserl's conception, is found in Simons 1987: chap. 8, pp. 290ff. Fine (1995) explores an alternative, nonmodal conception of dependence. Thomasson (1998) develops a succinct model of a basically Husserlian conception of dependence, refining distinctions formulated by Roman Ingarden and applying the model to fictional objects as dependent on authors' and readers' intentional acts. Another recent discussion is Lowe 1994. D. W. Smith (1989: chap. 6) sketches a ramified notion of dependence or ground, specifying different kinds of dependence (physical, psychological, etc.) and distinguishing kinds of dependence involved in direct awareness, or acquaintance, notably dependence on intentional content and different kinds of dependence on the context of one's experience. Simons (cited earlier) similarly distinguished different kinds of ontological dependence, including logical presupposition.

31. Barwise and Perry (1983: chap. 5) construct a logic of reports of perception and other attitudes, using a notion of "constraints" on meaning and other phenomena. There are some affinities between that notion of constraint and the notion of "background" under discussion here, but that is a topic for another day, as we are not here concerned with the logic or semantics of reports of intentional attitudes.

32. See Kim 1993, especially the essay "Supervenience as a Philosophical Concept."

33. See Burge 1979, 1982. Such externalism is critiqued in D. W. Smith 1991.

34. See Dretske 1981, 1995. Compare Gibson 1979.

35. The different roles of context and content are vivid in perception. There is normally a causal relation between the object perceived and the perceptual experience, but there is also an intentional relation between them, mediated by the content of the experience. The content prescribes the object as before

the subject and causing the experience. But the intentional relation is distinct from the causal relation: intentionality does not reduce to causal relatedness. This line of argument is detailed in D. W. Smith 1989.

36. Some of Jung's key writings on the notion of archetypes, from 1934 and 1935, are collected in *The Basic Writings of C. G. Jung* (1990).

References

Armstrong, David M. 1989. *Universals: An Opinionated Introduction*. Boulder: Westview Press.

Barwise, Jon, and John Perry. 1983. *Situations and Attitudes*. Cambridge, Mass.: MIT Press.

Burge, Tyler. 1979. "Individualism and the Mental." In P. French, T. Uehling, and H. Wettstein, eds., *Midwest Studies in Philosophy*. Minneapolis: University of Minnesota Press.

⸺ 1982. "Other Bodies." In Andrew Woodfield, ed., *Thought and Object*, pp. 97–120. Oxford: Clarendon Press.

Castañeda, Hector-Neri. 1975. *Thinking and Doing*. Dordrecht: D. Reidel.

Dennett, Daniel C. 1991. *Consciousness Explained*. Boston: Little, Brown.

Dretske, Fred. 1981. *Knowledge and the Flow of Information*. Cambridge, Mass.: MIT Press.

⸺ 1995. *Naturalizing the Mind*. Cambridge, Mass.: MIT Press.

Dreyfus, Hubert L., ed. 1982. *Husserl, Intentionality and Cognitive Science*. Cambridge, Mass.: MIT Press.

⸺ 1991. *Being-in-the-World*. Cambridge, Mass.: MIT Press.

Fine, Kit. 1995. "Ontological Dependence." Paper presented at the meeting of the Aristotelian Society, Birkbeck College, London, 5 June.

Føllesdal, Dagfinn. 1982. "Husserl's Notion of Noema." In Dreyfus 1982. First published in *Journal of Philosophy* (1969).

⸺ 1990. "The *Lebenswelt* in Husserl." In Leila Haaparanta, Martin Kusch, and Ilkka Niiniluoto, eds., *Language, Knowledge, and Intentionality: Perspectives on the Philosophy of Jaakko Hintikka*, pp. 123–43. *Acta Philosophica Fennica* (Helsinki) 49.

⸺ 1992. "Gödel and Husserl." Paper presented at a conference on The Development of the Foundations of Mathematics, Boston.

Gibson, James Jerome. 1979. *The Ecological Approach to Visual Perception*. Boston: Houghton Mifflin.

Heidegger, Martin. 1969. *The Essence of Reasons*. Translated by Terrence Malick. Evanston, Ill.: Northwestern University Press. German original, 1929.

⸺ 1991. *The Principle of Reason*. Translated by Reginald Lilly. Bloomington: Indiana University Press. German original, 1957.

⸺ 1993. *Basic Concepts*. Translated by Gary E. Aylesworth. Bloomington: Indiana University Press. German original, 1981, from the text of a lecture course given in winter 1941.

Hintikka, Merrill B., and Jaakko Hintikka. 1986. *Investigating Wittgenstein*. London: Basil Blackwell.

Husserl, Edmund. 1962. *Ideas Pertaining to a Pure Phenomenology and to a Phenomeno-logical Philosophy, First Book: General Introduction to Phenomenology.* Translated by W. R. Boyce Gibson. New York: Collier Books. German original, 1913.

1970a. *Logical Investigations.* Vols. 1 and 2. Translated by J. N. Findlay from the revised, second German edition. London: Routledge and Kegan Paul. New edition, edited with an introduction by Dermot Moran, and with a preface by Michael Dummett. London: Routledge, 2001. German original, 1900–1, revised 1913 (Prolegomena and Investigations I–V), 1920 (Investigation VI).

1970b. *The Crisis of European Sciences and Transcendental Phenomenology.* Translated by David Carr. Evanston, Ill.: Northwestern University Press. German original published in 1954 and edited by Walter Biemel, from materials written in 1935–38.

1989. *Ideas Pertaining to a Pure Phenomenology and to a Phenomenological Philosophy, Second Book: Studies in the Phenomenology of Constitution.* Translated by Richard Rojcewicz and André Schuwer. Dordrecht: Kluwer Academic Publishers. Original manuscript dating from 1912, posthumously published in German in 1952.

Ingarden, Roman. 1964. *Time and Modes of Being.* Translated by Helen R. Michejda. Springfield, Ill.: Charles C. Thomas. From excerpts of *The Controversy over the Existence of the World* (in Polish, 1946–47).

1989. *The Ontology of the Work of Art.* Translated by Raymond Meyer with John T. Goldthwait. Athens: Ohio University Press. German original, 1961.

Johansson, Ingvar. 1989. *Ontological Investigations.* London: Routledge.

Jung, Carl Gustav, 1990. *The Basic Writings of C. G. Jung.* Translated by R. F. C. Hull. Edited and with an introduction by Violet de Laszlo. Princeton: Princeton University Press.

Kim, Jaegwon. 1993. *Supervenience and Mind.* Cambridge: Cambridge University Press.

Krauss, Ruth, and Maurice Sendak. 1952. *A Hole Is to Dig: A First Book of First Definitions.* Text by Krauss, pictures by Sendak. New York: Harper and Row.

Kusch, Martin. 1989. *Language as Calculus vs. Language as Universal Medium: A Study in Husserl, Heidegger and Gadamer.* Dordrecht: Kluwer Academic Publishers.

Lowe, E. J. 1994. "Ontological Dependency." *Philosophical Papers* 23 (1): 31–48.

Searle, John R. 1983. *Intentionality.* Cambridge: Cambridge University Press.

1990. "Collective Intentions and Actions." In Philip R. Cohen, Jerry Morgan, and Martha E. Pollack, eds., *Intentions in Communication,* pp. 401–15. Cambridge, Mass.: MIT Press.

1992. *The Rediscovery of the Mind.* Cambridge, Mass.: MIT Press.

1995. *The Construction of Social Reality.* New York: Free Press, Simon and Schuster.

Sellars, Wilfrid. 1963. "Philosophy and the Scientific Image of Man." In *Science, Perception and Reality,* pp. 1–40. London: Routledge and Kegan Paul; New York: Humanities Press.

Simons, Peter M. 1987. *Parts.* Oxford: Oxford University Press.

Simons, Peter M., and David W. Smith. 1993. "The Philosophical Foundations of PACIS." Paper presented at the 16th International Ludwig Wittgenstein Symposium, Kirchberg-am-Wechsel, Austria, 19 August.

Smith, Barry, ed. 1982. *Parts and Moments*. Munich: Philosophia Verlag.

    1995. "Common Sense." In Barry Smith and David Woodruff Smith, eds., *The Cambridge Companion to Husserl*. Cambridge: Cambridge University Press.

Smith, David Woodruff. 1989. *The Circle of Acquaintance*. Dordrecht: Kluwer Academic Publishers.

    1991. "Thoughts." *Philosophical Papers* 19 (3): 163–89.

    1992. "Consciousness in Action." *Synthese* 90: 119–43.

    1995. "Mind and Body." In Barry Smith and David Woodruff Smith, eds., *The Cambridge Companion to Husserl*. Cambridge: Cambridge University Press.

Smith, David Woodruff, and Ronald McIntyre. 1982. *Husserl and Intentionality: A Study of Mind, Meaning, and Language*. Dordrecht: D. Reidel.

Thomasson, Amie. 1998. *Fiction and Metaphysics*. New York: Cambridge University Press.

Tougas, Joseph. 1996. "Hertz und Wittgenstein: Zum historischen Hintergrund des Tractatus." *Conceptus* 29, 75: 205–28.

Wittgenstein, Ludwig. 1972. *On Certainty*. Edited by G. E. M. Anscombe and G. H. von Wright. Translated by Denis Paul and G. E. M. Anscombe. New York: Harper Torchbooks, Harper and Row. First published by Basil Blackwell, 1969, from notebooks written in German during 1949–51.

    1992. *Tractatus Logico-Philosophicus*. Translated by D. F. Pears and B. F. McGuinness. Atlantic Highlands, N.J.: Humanities Press International. This translation first published in 1961. German original, 1921.

6

Intentionality Naturalized?

Abstract. Here I outline an ontology of diverse *categories* that define a place for consciousness and intentionality in the world of "nature." Intentionality is "naturalized" only in that way, without reducing consciousness or its intentionality to a causal or computational process. The basic background assumption is that there is one world ordered and unified as "nature." The world includes us, our conscious intentional experiences, physical objects, and social organizations. These things belong to diverse "material" categories such as Body, Mind, and Culture, which are ordered by diverse "formal" categories such as Individual, Quality, and State of Affairs. The world is unified by the systematic ways in which these categories interact. Among the formal categories of the world, I claim, are Intentionality and Dependence. If these are distinct formal categories, it is a mistake to identify intentionality with a structure of causation (dependence) realized in a brain or computer. We need instead a more sensitive system of categories.

Segue. In "Consciousness in Action" we showed that the experience of everyday action includes a sense of dependence on embodiment. Then in "Background Ideas" we argued that familiar forms of intentional experience depend on ideas extant in one's background culture. Now, consciousness – in thought, perception, and action – is part of nature, and so is human culture. However, to articulate the mind's place in the world of nature, we need a broad and fundamental ontological framework. Yet the type of naturalism assumed in today's philosophy of mind *cum* cognitive science cannot

Various participants in the Bordeaux conference where the material in this chapter was first presented offered helpful comments, responding to my presentation of these ideas there. I thank Charles W. Dement as well, for discussion of related issues and for comments on the penultimate draft of this essay. My thanks also to Jeffrey Barret for comments on the penultimate draft and to Dagfinn Føllesdal, Ronald McIntyre, Peter Simons, Andrew Cross, and Martin Schwab for helpful discussion of several related issues.

make room for conscious, embodied, encultured intentional activities. In this essay we look toward a wider and more fundamental ontology. We turn thus to issues of basic ontology to be explored further in subsequent essays.

With all its eyes the natural world looks out into the Open. Only our eyes are turned backward, and surround plant, animal, child like traps, as they emerge into their freedom.

Rainer Maria Rilke

Introduction

Intentionality in Nature

We must open up the conception of nature and naturalism au courant in the philosophy of cognitive science. Consciousness and its intentionality are part of nature. But their essence is not exhausted by physical composition or causal role or neural function or computation (classical or connectionist) – as important as these be to their implementation. We must widen "naturalism" accordingly so that our theory of mind and cognition incorporates the characters of consciousness and intentionality that are hard-earned results of phenomenology and ontology.

To that end I outline here an ontology that distinguishes diverse *categories* of the world and carves a place for consciousness and intentionality in the world of nature. Intentionality is categorized within the world of nature, and "naturalized" in that way, without reducing consciousness or its intentionality to a causal or computational process along the lines envisioned by current cognitive science.

The guiding assumption, which I call *unionism*, holds that there is but one world, ordered and unified as "nature." That world includes us, our conscious intentional experiences, and a host of other things, from rocks, trees, and bees to families, symphonies, and governments. These things belong to diverse substantive or "material" categories such as Body, Mind, and Culture, which are ordered by diverse "formal" categories such as Individual, Quality, and State of Affairs. The world is unified by the systematic ways in which formal categories interweave and govern material categories.

This proposed unity does not preclude diversity of categories (think of the classifications in biology, the table of elements in chemistry, the fundamental forces and particles in physics). Yet diversity does not require disjoint realms of beings delimiting, say, bodies, minds, and cultures (à la dualism, materialism, idealism, historicism). In this way the world enjoys categorial complexity without substance dualism. From quarks to quasars, from consciousness to volition to cultural institutions, there is ontological complexity, but all within this one world.

Among the formal categories of the world, I shall claim, are Intentionality and Dependence. But if these are distinct formal categories, it is a mistake in fundamental ontology to identify intentionality – or consciousness itself – with a structure of causation, or dependence, realized in a brain or computer. The life of mind is more complicated than that, categorially.[1]

Naturalism in Phenomenology and Cognitive Science

Some ninety years ago Edmund Husserl developed a philosophical account of consciousness and cognition in *Ideen* (1912–13) on the heels of his *Logische Untersuchungen* (1900–1). Husserl held that "intentionality" (Husserl coined the term) is the central feature of consciousness. His account of intentionality – including cognition in perception and judgment – was the foundation of "phenomenology," Husserl's new science of consciousness. With Husserl the theory of intentionality and intentional content thus came into its own, after a long prehistory.

Some twenty years ago, at the heart of the emerging discipline of "cognitive science," Jerry Fodor outlined a philosophical account of cognition in *The Language of Thought* (1975). Fodor held that "mental representation" – what Husserl called "intentionality" – is the central activity of mind and analyzed its structure as symbolic computation in "the language of thought." More recently, Fodor and foes have debated whether this computation is "connectionist" in form, implemented by neural networks in the human brain, whereas Fred Dretske, since *Knowledge and the Flow of Information* (1981), has put the emphasis rather on the flow of physical "information," between environment and organism, in perception and knowledge.

But Husserl and Fodor et alii take opposite positions on naturalism. Expressing today's conventional wisdom, Fodor in *The Elm and the Expert* (1994) says any "serious psychology" must be naturalistic. In kindred spirit Fred Dretske in *Naturalizing the Mind* (1995) seeks to show how consciousness might be captured in naturalistic terms. Husserl, however, argued that any "rigorous science" of consciousness must reject naturalism. Of course, everything depends on how naturalism is defined. If intentionality is naturalized in the way of Fodor et alii, then I stand with Husserl against "naturalization." But if naturalism is defined in the way

I shall propose, as unionism in a system of ontological categories, then I accept a "naturalized" intentionality, and so I think could Husserl in today's debate as well.

In *The Elm and the Expert* Fodor updates his account of mental representation, now adopting Husserl's term "intentionality" and drawing what he thinks are consequences of naturalizing the theory of mind and intentionality. In *Naturalizing the Mind* Dretske proposes to explicate the crucial features of consciousness, including intentionality, qualia, and introspection, in terms of external relations our experiences have to relevant objects in the environment, relations consisting in the causal flow of physical "information." Where Fodor, Dretske, and others would outline a theory of mind within the bounds of a naturalism stressing computation and causation, I want inversely to outline a form of naturalism within the bounds of a theory of mind (inter alia) stressing its place in the categories of a phenomenological ontology.[2]

The term "naturalism" outlives its usefulness, however, if it merely connotes the honorific attitude, "Be scientific." My aim here is to get beyond such labels and dig into what would be required of an ontology adequate to the phenomenology of consciousness and its intentionality.

Outlines of a Phenomenological Ontology

Ontological Categories

Systematic ontology begins by distinguishing fundmental kinds or *categories* of things in the world. Plato distinguished forms and particulars that exemplify them (to various degrees of imperfection). Aristotle, master classifier and protobiologist, framed a more down-to-earth ontology of "categories" (launching this term of art in philosophy: *kategoremata*, literally "predicates"). Aristotle distinguished the following categories:

> Substance, including Primary Substance (Individual Thing) and Secondary Substance (Species), Quantity (Discrete or Continuous), Quality (Condition, Capacity, Affective Quality, or Shape), Relative (better Relation), Where, When, Position (Posture), Having, Doing (Action), Being Affected (Being Acted Upon).

If Plato's forms reside in a heaven far above the world of particulars, Aristotle's categories – starting with Substance – are meant to structure things in nature. Thus Aristotle went on to distinguish matter and form within the constitution of a primary substance.

For our purposes let us simplify and update Aristotle's system of categories, as follows:

Individual, Species, Quality, Relation, Location, Quantity, Intentionality.

Species include natural kinds and, if you like, artificial kinds. Relations form their own category (by contrast, many philosophers sought to reduce relations to monadic properties). Space and time fall under Location. Quantity leads into other mathematical structures (details to be specified). Intentionality – following Brentano, Husserl, et alii – covers not only purposeful action (and, inversely, being acted upon) but perception, imagination, thought, and other processes (significantly widening Aristotle's niche for acting and being acted upon).

We consider increasingly complex systems of categories, using this modified Aristotelian scheme as a prototype and foil for further reflections. The issue, as we proceed, is the niche of mind in a world framed by ontological categories. If Brentano was right that intentionality is the distinguishing mark of the mental, then of course mental phenomena would simply fall under Intentionality in the above scheme. But life is more complicated than that, we shall see.

Classical nominalism began in Ockham with the goal of having only the category Particular, or Individual, what Aristotle called "Primary Substance": all else becomes mere ways of talking or thinking (for Ockham, thinking in a mental language). While I cannot argue the case here, I think nominalism – better named particularism – is orthogonal to naturalism. Our theory of nature assumes the categories Species, Spatiotemporal Location, and more. But whether these categories could be somehow eliminated, or reduced to the category Individual, is a further issue and would survive the denial of naturalism – as we saw in Berkeley's ontology of idealism (which would recognize only minds and ideas, two kinds of particular).

Aristotle conceived his categories as the most general yet disjunct kinds of things there are. This would seem to mean the categories themselves fall under the category Species. But that is not right. The categories are not *summa genera*, species at the highest level in the genus-species hierarchy: they are not species at all. Dog is a species of Mammal, but Species, Relation, and the like are not species: they are *categories*. Thus, a more systematic account of categories is needed, along the lines charted by Husserl.

Formal and Material Categories

Husserl drew a distinction (with a nod to Leibniz) between formal and material ontology, thereby distinguishing "formal" and "material" categories.[3] While the distinction is not easy to draw in general terms, it becomes clear in particular systems of ontology, where it has sharp consequences for the ontology of mind. When we begin to think in such terms, as we shall see, the reduction of intentionality to causal-computational process looks like a serious category mistake.

According to Husserl, *material* ontology recounts material categories of things according to their specific characters or "essence," whereas *formal* ontology recounts formal categories that apply to things of any material category. If you will, formal ontology posits *forms* or structures of things in the world, and material ontology posits "matters" or domains to which "forms" may apply. In this sense formal categories are topic-neutral. Indeed, Husserl sometimes spoke of the ideal of a *mathesis universalis* (Leibniz's term), suggesting that formal ontological structure is to be described ultimately in a unified mathematical framework. However, a formal ontology is formal because it describes ontological form, not because it may ideally be expressed in a symbolic or mathematical language, and not because forms are identified with mathematical entities such as numbers and sets.

For instance, if Descartes were right that minds and bodies belong to two distinct realms, both realms would still be characterized by the formal distinction between substances and their modes (things and their properties). Applying Husserl's distinction to Descartes's ontology, we would say that Mind and Body are two material categories, while Substance and Mode are two formal categories that apply to each of these material categories. Thus the mental divides into mental substances (minds) and mental modes (modes of thinking), whereas the bodily divides into corporeal substances (bodies) and corporeal modes (modes of extension).

Or consider Husserl's ontology.[4] He recognized three material categories: Nature, Culture, and Consciousness; and several formal categories including: Individual, Species, and State of Affairs. In Husserl's system, the material categories are the highest genera under which species can fall, and the formal categories are forms that apply within each material category, so that an individual's belonging to a species (which falls under the genus Nature or Culture or Consciousness) forms a state of affairs. Thus each of the "regions" Nature, Culture, and Consciousness is structured by the "forms" Individual, Species, and State of Affairs.

In Husserl's ontology the category State of Affairs (*Sachverhalt*) is formal and has a "syntactic" structure combining individuals and essences. Wittgenstein adopted a similar view in *Tractatus Logico-Philosophicus* (1921). According to Wittgenstein, the *form* of the proposition "*aRb*" is shared by the state of affairs (*Sachverhalt*) that objects *a* and *b* stand in relation R – and the form of an object is the possibility of its occurrence in various states of affairs. Wittgenstein's famous Tractarian ontology is thus a formal ontology in Husserl's sense and indeed is centered on the ontological form State of Affairs. Thus, the Tractarian forms Object, Relation, and State of Affairs would apply to any domain – mind, nature, culture, or whatever material ontology posits.

Issues in philosophy of mind take their place in such an ontology of formal and material categories. Thus, for Husserl, unlike Descartes, the material categories do not define disjoint domains of "substances," but rather different ranges of essence, or species. Accordingly, for Husserl, the same individual can belong to different species (have different essences) under different material categories. Notably, "I" – this one individual – am a physical organism in Nature, a person in Culture, and an ego in Consciousness. More precisely, the individual I includes these three different aspects or "moments" (particularized properties) that fall respectively under the material categories Nature, Culture, and Consciousness. And there are *dependencies* among these moments, as my state of consciousness depends on what my brain is doing and what my culture has done. Consequently, Dependence is a *formal* feature of entities falling under these three *material* categories, and holding even between entities in different material categories.

Furthermore, in Husserl's scheme, *intentionality* (being-conscious-of-something as well as being-an-"I") belongs to animal organisms in Nature, to persons in Culture, and to "pure" egos in Consciousness. It follows that Intentionality is a *formal* feature of entities bearing these three different *material* essences. Husserl does not explicitly classify intentionality as a formal feature, but I later consider reasons why it should be so classified.

For Husserl, intentionality is "pure" in consciousness but also "bound" into nature and culture. Of course, intentionality is realized in nature not in rocks but only in certain physical systems including animal organisms. And intentionality is realized in culture not in buildings or machines but only in the persons who build and use them. In this view intentionality is a formal property, in the sense that it applies to things in different "material" domains. It is not, however, "formal" in the sense of a syntactic form of physical symbols. To say with Husserl that intentionality is a formal feature in a formal ontology is not to endorse (with Fodor et al.) the computational theory of intentional content as a syntactic form of physical symbols processed by a brain or a computing machine. Quite the contrary, as we shall see.[5] Ontological form is not form of symbols but form of entities – which may be represented by symbols in virtue of *their* form (the structure of such representation being the purview of semantics).

In logic we assume a distinction between logical forms and the expressions to which they apply in forming more complex expressions. For example, the logical form "*p* and *q*" applies to different sentences that may fill the positions of *p* and *q*; or the logical form "the *F*" applies to definite descriptions that may be formed from different predicates "*F*." The distinction between formal and material ontological categories is like a projection of this logical distinction onto the world – or vice versa, *pace* Wittgenstein! But the point is that this distinction applies to structures of the world as opposed to structures of language. In the spirit of W. V. Quine, we might hold that our language and its logic go hand in hand with our ontology, so that both are subject to revision in light of advancing theoretical inquiry. Nonetheless, our ontology is a further thing, and in the spirit of Husserl formal and material categories together structure the world.

Within the structure of language, Quine distinguished, as had Bolzano, between logical forms and the expressions to which they apply. The correct logical forms, for Quine, are those in natural language as regimented by first-order logic. According to Quine, the ontological commitments of our language are expressed only by bound variables, or pronouns: only expressions of that logical form genuinely posit entities; only through so positing entities do we "reify" and individuate objects.[6] So Quine's position yields in effect a *formal* nominalism, holding that all entities in the world have the form specified by the logical form of variables, thus recognizing only the formal category Individual. Husserl would say other forms of language also carry ontological commitment, as predicates may refer to essences or species and sentences to states of affairs. So Husserl's formal ontology is wider, recognizing the formal categories of Essence or Species, State of Affairs, and Individual – which involve other formal features including Dependence and Intentionality.

The formal-material distinction is not the analytic-synthetic distinction. Quine argued against a sharp distinction between analytic and synthetic statements, the former true by virtue of meaning alone. Yet he held to a distinction between logical and nonlogical statements, the former true by virtue of logical form alone. Where Quine applied the logical-nonlogical distinction to expressions in a language, Husserl applied the formal-material distinction to objects in the world and to grammatical categories in language as a special case. For Quine our ontology is projected from our language, from our bound variables, and logical form is a matter of convention. (The category of terms may be hard-wired into human language ability; thus Quine once remarked, "Man is a body-minded animal.") For Husserl our ontology is projected from our intentional experience, expressed or expressible in language, and that ontology applies to language and intentional experience and content as well as to other kinds of objects in the world.

Philosophers have often thought of the logical or conceptual as prior to the empirical, and along these lines one might be tempted to think of formal ontology as prior to material ontology. However, we must instead posit formal categories together with material categories, in a unified ontology. For in practice we abstract formal features from material features of things, and if we revise our material categories sufficiently, we may revise our formal categories that govern them. Similarly, Quine holds that our logical and mathematical idioms are most central in the language expressing our web of beliefs, yet even they are subject to revision.[7] Indeed, some say quantum mechanics may require a different "logic" than everyday affairs. Moving to the level of ontology, our formal categories are more central and less variable than our material categories, yet the formal and material work together, and our ontology as a whole is subject to revision as need be.

The Ontology of Nature

As we distinguish formal and material categories, let us look toward a categorial ontology of nature. Think of the material categories implicit in physics and the formal categories that might govern these. As a model (subject to revision) contemplate the following categorial ontology of the physical world, the domain of physics:

Formal Categories: Individual, Species, Quality, Relation, Location, Quantity, Dependence, Intentionality, State of Affairs.

Physical Material Categories: Body, Wave, Mass, Force, Space-Time, Gravitation, Electromagnetism, Quantum Field, Wave Particle.

What I want to stress is the *architecture* or *systematics* of such an ontology, where the nature of things physical is structured by the interaction of formal and material categories. On such an ontology, the world has a systematic unity that consists in the way entities under material categories are governed by formal categories that weave together in an ordered way. In the ancient idiom "nature" (*phusis*) meant the order of things, and ontological systematics focuses on such order, framed here by a system of formal and material categories.[8]

These formal categories above are those recounted earlier as an updated Aristotelian scheme but supplemented here with the categories State of Affairs and Dependence. The category State of Affairs serves to bind together entities of other formal categories, as states of affairs are formed by individuals of various material categories having appropriate qualities, relations, locations, and so on. Thus, in its formal structure the world is a world of states of affairs (following Husserl, Wittgenstein, et al.).[9] But the world is also bound together by dependencies. In this ontology Dependence is posited as a formal category that may govern material causal relations defined in terms of physical force, wave activity, and the like. Thus, *A* depends on *B* just in case *A* could not exist or occur unless *B* does; this structure is realized in causation between physical things or events in virtue of the mass, force, or other physical entities or features that they involve. (Aristotle assumed such a notion in defining qualities of substances, but it was Husserl who developed the ontology of dependence in a detailed way.)[10] Note that Intentionality is posited as a distinct *formal* category, a point to which we return.

In the Cartesian and Husserlian ontologies sketched earlier, the formal categories apply in the same way to entities in each material category. In the present ontology, however, the relation between formal and material categories is more complicated, with certain formal categories applying to entities in certain material categories. Thus, a body is an individual with a mass in a gravitational field, but an electromagnetic wave is an individual with a certain mode of propagation in an electromagnetic field. States of affairs are formed in various ways from formal categories: as individuals belonging to species or having qualities or standing in relations, or entities depending on other entities.[11] The laws of physics, according to such an ontology, describe intervolved formal and material entities or features. For example, "$f = ma$" describes a relation among quantities correlated with a certain force, the mass of a certain body, and the rate of change of its location in space-time. Certain of the entities involved are "formal" and others are "material."

The categories of physics are subject to raging controversies, specifically, in the interpretation of quantum mechanics. Under material categories, are there waves only, particles being illusions brought on by interference of observation? Or are there particles only, determinate elements in eigenstates following collapse of the wave function? Or are there only superpositions forming wavicles, neither particles nor waves in themselves but only relative to observation or the absence thereof?

It remains unclear exactly which ontological categories are best suited to current physical theory. The point here, however, is not to settle such issues, but to indicate the relevance of the distinction between formal and material categories – and, more generally, the importance of "systematics" in ontology. However these controversies in the foundations of physics turn out, would the formal distinction between Individual, Quality, and Relation be abandoned? Is this much not presupposed by the most basic structures of mathematics, recognizing things and functions (with or without a strongly realist interpretation of mathematics)? It is hard to see how the different versions of quantum mechanics, or other parts of physics, could do without this formal distinction.[12]

In this category scheme, Intentionality is posited as a formal category because it may be required by the material categories presupposed by quantum physics. For on the Copenhagen interpretation of quantum mechanics, observation causes the collapse of the wave function in a quantum system. But if observation is an intentional event (as it is), and intentionality has a distinctive formal structure (as it does), then quantum mechanics (on that interpretation) presupposes the formal category Intentionality. Thus, the ontology of the physical world may require the category Intentionality even before we look at the categorial structure of mind per se.

The Ontology of Mind in Nature

Now let us turn to the ontology of mind. An adequate ontology must be *phenomenological*, in the sense that it recognizes structures of experience, notably consciousness and intentionality, as well as structures of stones, gravity, quantum states, and so on. On the present approach, a phenomenological ontology would be articulated within a scheme of formal and material categories.

For the sake of argument, assume the formal categories given previously. What material categories should an ontology of mind assume? They might be the same as those given for physical entities: this would be a form of physicalism, but the characters of consciousness and intentionality would have to take their place somehow in the category scheme. Or in a plural-aspect scheme like Husserl's, the formal categories would apply to, say, the material categories of Biomatter, Culture, and Consciousness. Each of these material categories would be characterized by diverse properties, but each would be governed by the formal categories of Individual, Species, and the like.

Rather than begin with the material categories of physics, however, our theory of mind and cognition must begin with categories of things in the everyday world of common sense – what Husserl called the "life world," the midworld range of things we live among between quarks and the cosmos.[13] We do not begin our ontology of the world around us by dealing with bosons and black holes, or neurons and neural nets, abstracting so far from our familiar concerns that we no longer know where we fit into the world. Rather, we encounter and categorize mental or cognitive activities *in everyday life* – prior to our theorizing in physics or neuroscience, which is constrained by our everyday experience and answers to it in our scientific practice of observing and theorizing. Indeed, our quantum physics and neurophysics must ultimately jibe with everyday experience so as to recognize and account for everyday affairs, including our conscious activities of theorizing.

Consider then the following categorial ontology of the everyday world, structuring the world as we know it in everyday life:

Formal Categories: Individual, Species, Quality, Relation, Location, Quantity, Dependence, Intentionality, State of Affairs.

Everyday Material Categories: Object, Event, Place, Time, Plant, Animal, Human Being, Mind, Action, Practice, Artifact, Institution.

Here, an object is an individual that persists relatively unchanged, an event is an individual that transpires relatively quickly in time, and so forth. The formal and material categories interweave so that there is ontological complexity in the world. But there are no separate domains of mental and physical substances or simple division of properties into mental and physical.

In this ontology there is one world that includes mind and culture amid "natural" objects or events. Mind takes its place in the order of the world, along with stones, trees, and owls, as well as cities, schools, and stories. A *mental activity* such as a thought, perception, or action is an intentional event: it is categorized materially under Event and formally under Intentionality. In closer detail, a mental activity is experienced by a subject (human or animal), carries an intentional content (a concept, proposition, etc.), and, if veridical, is directed to an object (the object of thought). There may be collective as well as individual intentional activities, such as collective beliefs or group intentions. The salient mental events are conscious: we *experience* these, and build our ontology around them among other familiar things. Now, *cultural* entities also take their place in the world. A hammer is an artifact: categorized formally under Individual and materially under Artifact. A political party is an institution, classified under Individual and Institution; a song is something we write and perform, an artifact, and so forth. *Cognition,* then, consists in a mental activity of perceiving or thinking being intentionally directed to an appropriate entity, which may be an object, an event, a plant, or a hammer. Notice that some features of mind or cognition are formal and some are material, instances of formal or material categories respectively. And notice, roughly, how the formal and material categories work together in these formulations.

These categories structure the things we deal with in everyday life, in our niche in the human world within the natural world. There are vital differences among the objects of everyday human life in the stone age, the agricultural age, the industrial age, and now the age of air travel, telecommunication, and computers. Yet the formal categories such as Individual or Species seem to apply to things under everyday material categories in any of these ages. And the everyday material categories given previously seem appropriate to different cultures of humans ancient and modern. Our categories are not written in stone (or iron or silicon), but we must revise them cautiously.

Unionism – the formal ontological backbone of an appropriate naturalism – requires that the world of nature is one world including quarks, quantum wave collapses, and black holes along with human beings and conscious intentional experiences and political scandals. Thus, the present ontology is unionist. As Husserl stressed, there must be dependencies of consciousness and human affairs on the basic physical states of things (and perhaps vice versa): dependencies among "moments" that instantiate the relevant categories, formal and material. From quantum mechanics to neuroscience there is much yet to learn about these dependencies. From categorial ontology, however, we draw an approach that systematizes the contributions to "formal" and "material" structures of the world. A unionist ontology would ultimately detail how the ontological structures of things physical, mental, and cultural are all extruded from a unified system of fundamental ontological categories, formal and material. The "correspondence problem" (borrowing the term from physics) is then how precisely to characterize relations among the physical, phenomenological, and cultural aspects of things – which are diverse aspects of things in this one world.[14]

Subcategories for Consciousness and Intentionality
Within the preceding ontology, under certain categories we may carve out subcategories that structure consciousness and intentionality more specifically, thus refining the formal and material categories that define mind.

Under the formal category Intentionality we specify formal subcategories that structure the intentionality of an experience. These are Subject, Experience, Content, Intention. Thus, the formal structure of intentionality is that of an *intention,* or "directedness," among subject, experience or "act of consciousness," content, and object. Content is a distinct formal category, since contents – images, concepts, propositions – are distinguished from species, relations, numbers, or quantities: they are formally their own kind of thing. And Content is a *formal* rather than material category because it is plausible, so far as we know, that the same content may be realized or invoked in different material categories, for example, if extraterrestrials have experiences that are realized differently than experiences in humans or animals. Our brains produce experiences through neural processes, but the extraterrestrial's system produces experiences, say, through gaseous propagation (in a life world like Jupiter's). This is not the functionalist claim that computation may be executed in different hardware or wetware or gasware. Rather, the claim is that thinking and other kinds of experience may be ontologically dependent on variable kinds of processes – processes that do not reduce to computation.

Cognition – in perception, thought or contemplation, and judgment – is formally classified as a kind of intention, or intentionality, not a structure of dependence. Yet our cognitive activities are materially analyzed as dependent on a causal flow involving the neural activity in our brains and its causal interaction with things in our environment. If this ontology is correct, then the reduction of intentionality or cognition to causal-computational process – as proposed in recent cognitive science – rests on a formal category mistake. It is not just that we happen to talk about intentionality and causality in different ways, even in different grammatical categories, or that we experience them in very different ways. Rather, the point is that we categorize them differently.

From our conscious mental activities, we may abstract material categories of their psychology and phenomenology. These are phenomenological material categories governed by the formal category

Intentionality. Thus, under the material category Mental Activity are subcategories specifying phenomenological types, namely: Perception, Thought, Emotion, Volition. These types of mental activity carry materially distinct types of content: perceptions, images and thoughts (propositions), feelings (and desires), and volitions (or intentions).

Conscious intentional experiences are thus categorized formally under Intentionality and materially under Mental Activity, with the foregoing subcategories. The category scheme in which they take their place then looks like this (ignoring further subcategories):

> *Formal Categories*: Individual, Species, Quality, Relation, Location, Quantity, Dependence, Intentionality, State of Affairs.
> *Subcategories* (under "Intentionality"): Subject, Experience, Content, Intention.

> *Everyday Material Categories*: Object, Event, Place, Time, Plant, Animal, Human Being, Mental Activity, Artifact, Institution.
> *Subcategories* (under "Mental Activity"): Perception, Thought, Emotion, Volition.

Mental life grades off from the conscious into lower levels of awareness and on into the unconscious – and into paranormal phenomena of split brains, split personalities, hallucinogenic experiences, and highly disciplined states of meditation. We begin our categorization, however, with conscious mental activities familiar in everyday life. Empirical investigation into the less familiar will extend our categories, perhaps modifying them as well.

Naturalism, Dualism, Supervenience, Functionalism

How does the categorial phenomenological ontology I have sketched differ from extant positions in philosophy of mind and cognitive science?

This ontology posits a diversity of categories in our one world. There are many kinds of things in the world and many basic categories, even different levels of categories (formal and material). Yet all things are tied into one world: "nature," if you will. This unionist view differs from extant forms of naturalism precisely in its account of categorial diversity.

Like all forms of naturalism, this ontology rejects substance dualism. It does not divide the world of concrete things into two domains, the mental things (minds, souls, spirits) and the physical things (bodies, from quarks to quasars). Rather, mind is part of nature, and so is culture: physical objects and events, mental activities, persons, societies, and cultural institutions all occur in this one world of nature. However, this categorial ontology rejects the simple identification of mind (and culture) with the workings of quarks, neurons, and so on. For it posits a more complex system of ontological categories.

Property dualism takes a step toward categorial diversity in positing two types of properties (as opposed to substances): physical properties (such as mass) and mental properties (such as intentionality). The categorial ontology presented here does not, however, divide properties into just these two categories. Instead, it posits a structured system of categories that divide the world into various groups of things, properties, and the like.

Supervenience theories hold that the mental "supervenes" on the physical – if we assume property dualism but deny substance dualism.[15] One form of supervenience is defined basically as the dependence of a mental event on physical events in a brain. Because dependence is one of the formal categories in the preceding ontology, supervenience can be placed in that ontology. However, dependence alone does not place mind in the world, as shown by the variety of categories we have adduced.

Functionalism takes a different step toward categorial diversity, distinguishing functional and physical properties of a system, and identifying mind with computational function of a neural system (at least in humans and other thinking animals on Earth). The functional-physical distinction can be seen as a formal distinction among types of properties. In the preceding categorial ontology, this distinction would be reconstructed in terms of a temporal sequence of events: a functional property is defined solely by a type of result in a sequence of events. However, this categorial ontology would reject the simple identification of mental event types with functional types. For there is more complexity in the categorial structure of mind (and culture) than function alone. To deny functionalism is not to deny function but to put it in its place.

The ontology above differs from these prominent approaches to philosophy of mind, then, in seeking a systematic account of the *categorial complexity* of the world, and mind's place therein. Recent forms of naturalism are explored from this perspective later, especially the causal and computational models of mind espoused by Fodor and Dretske.

Internalism and Externalism

Within the preceding categorial ontology, what are we to make of the debate between internalism and externalism in philosophy of mind?

On the *internalist* view of intentionality, what makes a mental activity intentional, or directed toward its object, is the intentional content held in mind, where that content is internal to the mental act. By contrast, on the *externalist* view, what makes a mental act intentional is context, either its physical causal history or its cultural circumstance, whence intentional content must be defined in terms of the act's contextual relations. The classical view is internalist, as found in Husserl and defended recently by John Searle and (with a contextualist reformation) myself.[16] But the predominant attitude in recent cognitive science has been externalist, as expressed by Fodor and Dretske.[17]

As we work with the foregoing unionist ontology, however, it becomes clear that the opposition between internalism and externalism is false. A mental act has both a causal genesis and a phenomenological character. And these two aspects of the act belong to different ontological kinds: the former is an instance of the category Dependence, whereas the latter is an instance of the category Intentionality (and relevant subcategories). Our ontology must recognize both the genetic tree and the phenomenological character of a mental act, its "clade" and its "phenotype" (to borrow biological idiom) – and neither reduces to the other.

In the ontology, Intentionality and Dependence are distinct formal categories. They must be distinct because they have radically different ontological structures. Thus, the structure of intentionality is that of an experience (by a subject) intending an object via a content (a concept, proposition, image, etc.). By stark contrast, the structure of dependence is that of one thing's needing another in order to exist. The ontology of these structures is so obviously and fundamentally different that reducing the intentionality of an experience – or its content – to its causal dependence on other things, as externalists sometimes propose, is simply out of the question.

Intentionality as a Formal Category

Our Cartesian heritage shapes philosophy of mind as a choice among *material* categories of substance, so that we think we must say either "All is Body" or "All is Mind" or "All is divided between Body and Mind" – thus the lure of physicalism in the age of physics. Furthermore, the emphasis on empirical ("material") results of recent science – from quantum mechanics to cognitive neuroscience – leads us to overlook distinctions about mind that belong properly to formal rather than material ontology. Thus causal and externalist theories of mind in Fodor, Dretske, and others cathect on material patterns of causation, whereas functionalist and computationalist models of mind obsess over the marvels of computer engineering (the most "material" of results).

But systematic ontology – ontological systematics – leads us to think about mind in different ways, avoiding the usual suspects of substance dualism, idealism, physicalism, functionalism, and the like.

In an ontology that distinguishes formal and material categories, Intentionality belongs among the formal categories. To justify this assumption in detail would be a long story, arguing that the intentionality of an experience is a unique form differing from that of a quality, a relation, a dependence. To begin the story, the ontological structure of intentionality is reflected in the distinctive "intensional" logic of idioms such as "A thinks that . . ." – but we are talking about ontological structures of experiences in the world, not logical structures of our language about experiences.[18] Accordingly, we must distinguish the object of thought from the content, the idea or concept of the object, through which one thinks about it, through which it is "intended" by one's thinking. This structure of intention-via-content is formally unique. Thus, the ontological structure of an intention or intentionality is just plain different from that of a quality, a relation, a dependence, and so on. Hence, Intentionality, Quality, Dependence, and others are distinct formal categories.

It is natural to speak of the "intentional relation" between an experience and its object, yet this "relation" is so unusual that Intentionality and Relation deserve the status of distinct formal categories. The formal structure of an intention, or "intentional relation," is that of subject-act-content-object. Thus, a subject has an experience or act that has a certain content, which prescribes a certain object (if such object exists); intentionality is this directedness of act by subject through content toward object. Is this the structure of a *relation* among subject, act, content, and object? Either "yes" or "no" carries problems. If it is a relation, its number of relata varies: sometimes there is no object and it is a three-place relation; otherwise there is an object and it is a four-place relation. This is not normal for relations. As Brentano and Husserl observed, intention is "relation-like" if not a proper relation. If it is not a relation, then what? Then it has its own form, as described, and a distinct category.

The category Intentionality is formal because it may apply to materially different domains. The same form of intentionality, such as perceiving a window, may apply to very different experiences in a fish, a lizard, a parrot, or a cat, not to mention an extraterrestrial. The functionalist, with Fodor, would explain this by identifying mentation with computation, realizable in different physical systems and even through different algorithms. That is not my point. Without reducing mind to computation or function or causal role, we should see that Intentionality is a form that applies to different domains of mental activity just as Relation applies to different domains of individuals – whence intentionality is a formal, not material property of thoughts, perceptions, and the like.

What I want to stress is the role of intentionality in an ontology of mind among other things. If we recognize Intentionality as a *formal* category, we shall not be tempted to find the difference between consciousness and, say, electricity in two types of substance called "mind" and "matter." We shall instead articulate a unified world in which such different phenomena take their places.

Problematic Features of Consciousness and Intentionality

Three features of mental activity have been problematic in recent discussions in philosophy of mind *cum* cognitive science. These are intentionality, qualia, and consciousness. Within the ontology outlined earlier, where lie these crucial features of mind?

Consciousness is the great mystery of the day in cognitive science, yet the familiar turf of phenomenology. Now, part of the problem of consciousness is its formal structure and part is its material character. Three crucial aspects of consciousness are intentionality, reflexivity, and qualia.[19]

The *intentionality* of a mental activity, on the above ontology, is a formal feature instancing the formal category Intentionality. The structure of intention (subject-act-content-object) applies to both conscious and unconscious mental activities.

Now, part of the formal intentional structure of a conscious mental activity is its *reflexivity*. When I have a conscious experience, I am reflexively aware of my having the experience. This form may be specified in phenomenological description, articulating content, by saying: "In this very experience I now see that wriggling snake on the path before me."(See Chapter 3 for details.)

The prefix "in this very experience" specifies the reflexive structure or content that in part makes the experience conscious. That reflexivity makes it, as Sartre stressed, a "consciousness (of) itself": not a separate act of introspection or reflection on the experience, but a prereflective awareness of the experience as it transpires. For Sartre, indeed, this character is definitive or constitutive of consciousness – an insight that cognitive science has yet to appreciate fully or to rediscover.

Quite a different feature of consciousness is the subjective character of an experience: its "phenomenal" or "qualitative" character, or qualia (especially its sensory characters) – in Thomas Nagel's idiom, "what it is like" to have that experience. What makes an experience conscious, I would urge, is the combination of qualitative character and reflexivity, which takes its place within the intentionality of the conscious experience.

To what category do qualia belong? Qualia are properties of an experience or act of consciousness. Their formal category is Quality. Their distinction from other qualities is a material distinction within the domain of conscious mental activities – falling under the material categories Perception, Thought, Emotion, Volition – to which the forms Quality and Intentionality apply. Specifically, they are distinct in type from qualities such as the spiking frequency of a neuron or its receptivity to certain neurotransmitters. The intentional form of an experience must make room for these qualities we call subjective.

Now, if Intentionality is a formal ontological category, and Mental Activity a material ontological category, then an event of cognition has the form of an intentional mental activity with a specific intentional content such as "I now see that snake on the path before me." The event may have the material essence of being implemented by a neural process in a human brain. Or if the event is a thought performed by a being on a distant planet, its material essence may be very different. Whatever its material essence, it may also have a form not of Intentionality but of Dependence – its causal role, which is a form of dependence on past and future events. But because Intentionality and Dependence are different formal categories, there is no temptation to reduce intentionality to causal-computational role. However, the formal category Intentionality does not apply to every material domain: rocks do not have intentionality, and neither do silicon chips. (This is not to beg the question of whether Intentionality reduces to Causation or Dependence. Remember, the formal properties of intentionality are not preserved by causal reduction.)

Finally, note that consciousness does not reduce to computation. Not all conscious activity is computation; feeling dizzy is not computing. Perhaps, as neuroscience suggests, a certain form of computation is part of the neural activity that implements every state of consciousness. But that is a different matter, a gray matter.

Critique of Naturalism

Fodor for Naturalism

Where classical phenomenology, in Husserl and his progeny, opposed something called "naturalism," today's cognitive science assumes naturalism as a given – the only option of good science – and seeks to "naturalize" mind and specifically intentionality.

Current attitudes toward naturalizing intentionality run the gamut, however. W. V. Quine once rejected the "science of intention" because it does not reduce to physical science and indeed requires a different logic than the rest of natural science. But he now gives our talk of intentional attitudes a place of honor alongside the language of natural science, and he says consciousness is something that must be explained.[20] Jerry Fodor assumes psychology must be intentionalistic – embracing intentionality in belief, desire, and so on – and also naturalistic; rejecting behaviorism and eliminativism along with idealism, he proposes to "naturalize" intentional content in the language of thought. Daniel Dennett endorses talk of intentionality, if only as a convenient "stance" toward what is really going on as the brain crunches data; his "explanation" of consciousness explains it (away) as multiple drafts of code in the brain. Paul Churchland and Patricia Smith Churchland have sometimes denounced intentionality as a remnant of folk psychology, viewed as superstition rather than empirical wisdom; but they do not now dismiss consciousness itself, the crowning achievement of neural activity. David Armstrong identifies mental activity, including intentionality, with brain activity, and he analyzes consciousness as the brain's monitoring itself. Fred Dretske takes consciousness and intentionality seriously and analyzes their structure as a flow of causal "information" outside the head. Donald Davidson and Jaegwon Kim, meanwhile, have held that the mental and intentional "supervenes" on the physical activity of the brain – a ramified form of physicalism. And Francis Crick (Nobel Laureate biochemist) gives scientific credulity to consciousness and even the "soul" while declaring, "You're nothing but a pack of neurons."[21]

Fodor holds that "naturalizability" is required for any serious science, not just psychology or cognitive science. He defines *naturalism* in effect as the assumption that "everything that the sciences talk about is physical" (Fodor 1994: 5). This assumption boils down to *physicalism*, or *materialism*, the ontological view that everything is physical. When Quine proposed "naturalizing" psychology and epistemology, he urged following the methods of natural science (and expressing its results in first-order logic) – thereby securing a physicalist ontology, unless science leads elsewhere. He thus opposed beginning with a "first philosophy," whether a metaphysics (as in Aristotle) or an epistemology (as in Descartes) – or, we might add, a phenomenology (as in Husserl).[22] The naturalism in cognitive science today assumes both ontological and methodological positions, but I want to focus on the ontological.

Fodor thinks a naturalistic intentional psychology requires two doctrines:

1. *Causalism*, the view that intentional content (and also linguistic meaning) is "information," in the technical sense of a physical structure transmitted by a causal flow.
2. *Computationalism*, the view that thinking is computation, in the technical sense of computer science, that is, an algorithmically defined process in either a digital computing machine or a neural network in a brain.

(The labels "causalism" and "computationalism" I introduce here for convenience.) Causalism – championed by Dretske with Fodor following suit – is a variety of:

3. *Externalism*, the view that content or meaning is defined by its external relations to physical (or on some views social) relations with what lies "outside the head" and indeed outside the given intentional state.

And computationalism, of course, is a variety of:

4. *Functionalism*, the view that thinking is wholly defined by its function in a physical system as it interacts causally with other processes of thinking and ultimately with processes in the physical environment.

Classical computationalism also assumes:

5. *Syntactic formalism*, the view that the properties of thinking that count are formal, syntactic properties of physical symbols processed in brains or computing machines.

An alternative to classical computationalism is:

6. *Connectionism*, the view that the properties of thinking that count are vector distributions of relevant properties in dynamic networks of neurons in brains, which can be modeled or perhaps shared by computer systems with parallel distributed processing.

Some connectionists consider this view a paradigm shift away from classical computationalism (assuming syntactic formalism) and thus from the computer model of thought. Fodor, however, counters that the form of the computation, whether logicosyntactic or connectionist, is merely a matter of implementation.[23] And the implementation of computation, whether classical or connectionist, lies in causal processes. Hence causalism remains in force. Dretske, however, downplays computation inside the brain and stresses the causal flow of "information" outside the organism. So the project of naturalizing the mind, for Fodor and Dretske, is to show how intentionality and other features of mind can be accounted for in a way consistent with causalism, hence externalism – all in the service of naturalism, and ultimately physicalism.[24]

Unfortunately, neither causalism nor computationalism (nor connectionism) preserves the properties of consciousness and intentionality. The phenomenological character of consciousness (beginning with qualia) is not preserved, and intentionality is not simply identical with the structure of a causal process – whether realizing a connectionist or a classical computational architecture. Within philosophy of mind, in the cognitive science tradition, John Searle has argued clearly and effectively the inadequacy of the computationalist view: syntactic form alone does not yield intentional content or meaning; neither, we might add, does connectionist form. Consciousness and intentionality, Searle holds, are irreducibly subjective properties of mental states, realized, however, in our brains.[25] Earlier, within the phenomenological tradition, Husserl, Sartre, Merleau-Ponty, and others charted the properties of consciousness and intentionality in great detail, always distinguishing them from various natural, physical properties of human bodies.

The properties of intentionality do not fit into the causal-computational view Fodor prefers. Fodor's task in *The Elm and the Expert* is to bite the bullet and try to make his preferred view amenable to the basic features of intentionality. But Fodor recognizes that his favored naturalizing gets wrong the basic properties of intentionality. How does he justify going on with his approach? He says the problem cases – Frege cases and Twin cases – do not happen very often. That means his theory is wrong: they happen and his theory does not handle them. On consciousness he is silent, beyond saying it is not tractable, in his kind of research program.

Dretske in *Naturalizing the Mind* bites the bullet in closer detail, recasting the features of conscious experience as external, causal features. He treats intentionality as natural "indication": a perceptual state indicates its object as a speedometer indicates the speed of the automobile (all these things being physical entities). In effect Dretske builds in causal terms a replica of the structure of intentionality, and likewise for qualia and introspection. That gets the structure right but applies it to the wrong phenomena. By tracing the causal flow of physical "information" that gives rise to the important features of consciousness, intentionality, qualia, and introspection, Dretske shrewdly tracks the causal preconditions of such phenomena. (See Chapter 1 on substrate and Chapter 3 on levels of consciousness.) But he identifies the *physical structures* called "information" in information theory with the *intentional contents* called "information" in everyday talk about contents of thought or speech. Now, intentional "information" in one's thought may well depend on physical-causal "information" in one's environment and indeed in one's brain. But the two things have very different properties.

Although it may seem that causalism *cum* externalism, and/or computationalism *cum* functionalism, serve the needs of naturalism and current cognitive science, they do not fit with prior results of phenomenology and indeed ontology. Phenomenology, in concert with ontology, has detailed structures of conciousness and intentionality in well-defined arguments from well-defined observations. These are sound results achieved by basic principles of rational inquiry in good philosophy or good science (as detailed, notably, by Quine in his general epistemology).[26] Yet externalism – the broad assumption behind causalism, computationalism, and functionalism – does not jibe with a phenomenologically sensitive account of intentionality in perception and other indexical forms of intentionality, as I have argued in a detailed study of acquaintance.[27]

Fodor says, "I'm not going to fuss about ontology except where it matters" (1994: 3). Unfortunately, a full theory of intentionality and consciousness requires more ontology than is countenanced by Fodor and most philosophers of mind in cognitive science.

Nor do Fodor et alii fuss about phenomenology, even where it matters. Consciousness has become a hot topic in cognitive science ever since neuroscientists declared it worthy of hard science.[28] Yet the writings of Fodor et alii show little awareness of the extant results of phenomenology. Owen Flanagan's *Consciousness Reconsidered* is advertised with the proclamation that its author "examines more problems and topics associated with consciousness than any other philosopher since William James"; yet the book makes no mention of Husserl, Sartre, or any other phenomenologist, even though Flanagan wants to make room for phenomenology in cognitive science.[29] In a recent issue of *Scientific American* philosopher David Chalmers, writing in the new wave of consciousness-raising, says that conscious experience might yet be explained not by neuroscience but by "a new kind of theory" that gives new fundamental laws about subjective conscious experience.[30] Aptly he broaches the need to distinguish between subjective and objective aspects of "information." Of course, that "new" kind of theory is called phenomenology, following Husserl. Many of its laws are already in place about intentionality, content, sensation, temporal awareness, and consciousness of one's experience, one's body, oneself, and others.

If psychology is to be set within the bounds of naturalism, naturalism must be set within the bounds of an adequate phenomenology and ontology. To this end there is much to be learned from Husserl.

Husserl against Naturalism

Husserl opposed naturalism as an ontology that would reduce the essence of consciousness and intentionality to physical-biological properties. Today he would have opposed, with Searle, their reduction to causal or computational properties, as opposed to intrinsically subjective characters of experience. But Husserl's antinaturalism was part of a systematic phenomenological ontology whose details developed in *Ideas* I and especially *Ideas* II. Husserl opposed naturalism on methodological as well as ontological grounds, as he sought foundations for natural science in a "first philosophy" that began with reason and ultimately grounded all knowledge in phenomenology. Here I focus only on the ontological issues. (Other phenomenologists, from Heidegger to Sartre, have also rejected naturalism. Here I draw on my own reconstruction of Husserl's account of mind and body *sans* idealism.)[31]

Husserl posited three "regions" of being, which he called Nature, Spirit or Culture, and (Pure) Consciousness. These regions are not domains of individuals or events; they are species or "essences." They are the highest "material" essences or genera to which individuals or events may belong. Nature is characterized by physical properties including spatiotemporality, mass, causality, and life. (Under Nature Husserl distinguished the living from the merely physical and my "living body," or *Leib*, from my "physical body," or *Körper*.) Culture is characterized by persons in communities holding values and making history. And Consciousness is characterized by the subjectivity of experience and the central feature of intentionality. Importantly, one and the same individual or event may be characterized by all three essences. For instance, I am a biological organism of the species *Homo sapiens sapiens*, and thus under Nature; I am also a person teaching philosophy, and thus under Culture; and I am as well an ego experiencing this event of thinking, and thus under Consciousness.

More precisely, bringing in Husserl's ontology of dependent parts or "moments" (what some philosophers today call "tropes"): the same individual, I, has distinct *moments* that are particularized instances of the essences Consciousness, Culture, and Nature – say, my moments of thinking, being a teacher, and being over six feet tall. Moreover, there are relations of ontological dependence among such moments or tropes. Thus, my thinking depends upon – cannot occur without – certain activities in my brain and other activities in my culture. (There lies a specific doctrine of the "supervenience" of the mental on the physical and cultural. But that is another story.)

Because the same individual can fall under these three essences, Husserl is a monist about substances but a pluralist about essences. In today's terms his metaphysics is a plural-aspect view: a form of substance-monism and property-pluralism – or essence-trialism.

On Husserl's account, intentionality is a property of consciousness, or part of the essence Consciousness, whereas force and mass and energy and causation are properties of nature, or part of the essence Nature, whereas value and empathy and community are properties of culture, or part of the essence Culture. Husserl's "transcendentalism" boils down to just this distinction among three irreducibly distinct essences and the properties that characterize them. According to Husserl, the "transcendental" region of consciousness does not reduce to the "natural" region of physical causation or to the "cultural" region of community. For Husserl, these three ranges of essence are presented through three distinct ranges of sense or noema expressed in three distinct ranges of language about the intentional, the physical, and the cultural. There are thus three grades of "transcendental" involvement: at the levels of language, sense, and essence respectively.

Given Husserl's ontology, what is wrong with the causal-computational view of mind is that it wrongly identifies two "moments" of an event of thinking: those of its intentional character and its causal-neuro-computational role. These two moments may be parts of the same event, and even interdependent, but they are numerically distinct. Notice that it matters not whether the computational architecture is classical or connectionist.

In Husserl's idiom, "naturalism" is the ontology that reduces the essence of consciousness to the essence of "nature," reducing the property of intentionality to a form of spatiotemporal-causal connection between act and object of consciousness. Naturalism in that sense Husserl rejected, vehemently. There is however a different ontology of nature and consciousness, and thus a different conception of "naturalism," that fits well with principles of phenomenology and ontology along the lines articulated by Husserl and facilitates the results of cognitive neuroscience. The first section in this chapter sketched a model of such a phenomenological ontology.

"Nature" Reborn with Consciousness and Intentionality

The ancient meaning of "nature" survives today in the distinction between "nature" and the "nature of things." Originally "nature" (*phusis*) meant the nature of the world (*kosmos*), as the earliest Greek philosophers posited an order to things in the world – the primordial assumption that became science. Today "nature" means the world so ordered, and its order is detailed by the "natural" sciences.

Where nineteenth-century naturalism stressed the organic character of the world and our place in it, late twentieth-century naturalism has stressed the physical character of the world, including the biological, along with the methodology of the physical sciences. Yet "naturalism" in current philosophy of mind (in Fodor, Dretske, and many others) takes a much more specific form: embracing not only physicalism but computationalism, causalism, externalism. What we need now, to accommodate the results of both phenomenology and cognitive neuroscience, is a naturalism endowed with metaphysical imagination.

The proper conception of nature begins with the recognition that our intentional states of consciousness are part of nature, with the characteristics uncovered by phenomenological ontology. To see this, we must recognize two points not adequately addressed by Fodor et alii: first, that we *experience* states or events of consciousness, notably intentional states; and second, that events of consciousness enter into *relations of intentionality* as well as relations of causation (glossing intentionality and causality as relations). The first point is the crux of phenomenology, the second its cognate in ontology. We must recognize unconscious intentional states as well, but here let us focus on the conscious ones.

The fundamental insight of naturalism should be the view that there is but one world, ordered and unified as nature. This view I have dubbed unionism. The traditional term "monism" stakes a different claim, that there is only one substance or one kind of substance. Unionism, by contrast, says there is only one (actual) world.[32] Within the order of the world, however, there may be different categories of things, including fundamentally different kinds or properties. When we turn to mind, we find that consciousness and intentionality are fundamentally different in character from processes of causation and computation, and we should not shoehorn them into the wrong category. Consciousness and intentionality depend on and enter into causal processes, both inside and outside of the brain – notably in perception and action. And intentionality, at least in some forms, involves computation. But, while perfectly "natural" and part of this one world, consciousness and intentionality do not consist simply in causal or computational processes, as Fodor et alii hoped. As with life in general, ontology is complicated and phenomenology plays its role.

John Searle has rightly stressed that the properties of consciousness and intentionality are irreducibly subjective properties that nevertheless occur in nature in biological events in human and animal brains.[33] They are not merely syntactic properties (whatever that may mean for neural events as opposed to our linguistic symbols); indeed, syntax itself is in the eye or ear of the beholder, a matter of convention. Consciousness and intentionality have "internal" properties of representation, properties that do not reduce to causal sequence or the transmission of physical structures called "information." Their essence is not wholly "external," or "outside the head," in either causal or social relations.

First we must recognize the existence of consciousness and its phenomenological structure; this requires the practice of phenomenology (by whatever methodology works: let us not bog down over "*ēpochē*," "hermeneutics," etc.). Then we must get consciousness and intentionality into our ontology: we must place them properly, categorize them, in our ontology. To distinguish "subjective" properties of consciousness and intentionality from "objective" properties of stones and quarks is not to lapse into a dualism of mental and physical substance. It is rather to chart the distinctive features of things, recognizing a pluralism of categories of properties of various things – all in the one world in which we live.

The naturalizing of mind proposed by Fodor, Dretske, et alii would reduce intentionality and consciousness to the flow of physical "information" through a brain or of bits in a computer. What could motivate the picture of mind – our own experience – as nothing but bits in the void? I suspect two assumptions of perennial philosophy: there are only particles in the void; all else is illusion. The first assumption is a simplistic materialism, the second a foolish metaphysics. Both have ancient roots.

But consciousness is not an illusion.[34] And intentional structure in belief, desire, and the like is not (as the Churchlands used to suggest) a theoretical illusion, an old wives' tale from ancient folk theory. Conscious intentionality is, in fact, the medium of illusion: an illusion is the presentation of something in consciousness that does not exist (outside the "intention"). If we know anything at all, it is that we have conscious intentional experience of this and that. Such is the subject matter of phenomenology, and to deny it is folly. (See Chapter 2.) We cannot deny consciousness and its intentionality in the name of science, one of the great achievements of human consciousness, any more than we can deny human language, the medium in which science is carried out. What we can do instead is wonder how consciousness – along with language and thinking-in-language – is realized or implemented in the order of nature described by recent science.

And twentieth-century physics, its foundations still unsettled, strongly suggests that the world order is not that of particles in the void. In relativity theory, space-time itself has topological structure – it is not a "void." And in quantum mechanics, the things in quantum systems are not "particles," but things in wavelike states that may collapse into particle-like states. These eigenstates, for that matter (pun intended), may even be states of consciousness: that much is perfectly compatible with quantum physics. And on some interpretations of quantum mechanics, the collapse of the wave function is itself triggered by an observation, an act of consciousness – in which case consciousness is implicated in the structure of physical states below the level of what normally we call causal processes.

I do not mean to suggest that proponents of today's naturalism are ignorant of twentieth-century physics, not to mention biology and neuroscience. Quite the contrary. My point is that considerations of phenomenology and ontology along with physical and cognitive science lead to a very different view of the role of consciousness and intentionality vis-à-vis causal and computational processes.

I do mean to suggest that we all carry vague background assumptions that shape our theorizing in less than obvious ways. These attitudes are part of the "background" of intentionality, unearthed by phenomenological reflection on intentionality itself.[35] Perhaps they have the status of "memes," cultural ideas passed on like genes in the cultural realm.[36] And I do mean to suggest that the reigning form of naturalism is led by such background attitudes into the untenable reduction of the properties of consciousness and intentionality to causal and/or computational properties.

What we need instead is a wider ontology behind naturalism, a unionism that places mind in nature without losing the distinctive characters of consciousness and intentionality. This requires a sensitive metaphysics, an ontology with appropriate formal, material, and phenomenological categories – an ontology along the lines of that sketched here.

Conclusion: One World, with Consciousness and Intentionality

I have outlined a categorial phenomenological ontology according to which consciousness and its intentionality, reflexivity, and qualia are part of nature, yet carry their own distinctive characters. In such an ontology the world has a certain *unity*, articulated by formal and material categories, within which consciousness takes its place. Everything there is, from quarks to consciousness to governments and black holes, takes its place within this *one* world.[37]

The lesson for philosophy of mind, from phenomenology to cognitive science, is that "naturalism" looks dramatically different when we look to such a phenomenological ontology, wherein mind cannot be "naturalized" by reducing consciousness or intentionality to a causal or computational process. Yet only through such an ontology can we ever hope to understand the phenomena of consciousness and intentionality and their role in the natural order of things.

Notes

1. The term "material" is used here always to contrast with "formal," not to mean physical, or composed of matter-energy.
2. In 1993 the inaugural Jean Nicod lecture on philosophy of mind and cognitive science was given by Fodor in Paris, published in 1994 as *The Elm and the Expert*. In 1994 the Jean Nicod lecture was given by Fred Dretske in Paris, published in 1995 as *Naturalizing the Mind*. The aim of the conference in Bordeaux in October 1995, for which this chapter was written, was to assess the challenges of naturalization for principles of Husserlian phenomenology. I have used these books by Fodor and Dretske, appropriately, to set the stage for my project.
3. The following remarks about Husserl's ontology rest on my reconstruction thereof in D. W. Smith 1995.
4. Again see D. W. Smith 1995.
5. Compare McIntyre's argument (1986).
6. Quine's mature view of reification is found in his *Pursuit of Truth* (1992).
7. Compare Quine and Ullian 1978.
8. In biology systematics is the discipline of classifying living beings: distinguishing individuals, populations, species or taxa, and categories (Taxon, Family, etc.) – distinguished in terms of phena (observable qualities), synapomorphs (shared derived traits), and so on. Here I borrow the term "systematics" for the ontological discipline of classification of entities. The distinction between formal and material categories defines one type of ontological systematics. I focus on the formal-material distinction as a device that increases "systematic" complexity in fruitful ways, but ultimately the goal is to use ontological complexity to account for important distinctions concerning intentionality

and other features of ontological structure. My conception of ontological systematics derives from discussions with Charles W. Dement and Peter Simons,
in connection with the PACIS project at Ontek Corporation.

9. Wittgenstein's *Tractatus Logico-Philosophicus* (1921) posited such a structure
    in the world, at least the world as we find it, the world as represented in our
    logical systems. David Armstrong (1996) has argued recently for such an on-
    tology of states of affairs; see also Armstrong 1989. Prior to Wittgenstein's
    work, however, was Husserl's conception of states of affairs as a "formal"
    ontological category, in both his *Logical Investigations* (1900–1) and his *Ideas
    I* (1913).

10. The ontology of dependence is launched in the Third of Husserl's *Logical In-
    vestigations* (1900–1). Husserl's basic notion is reconstructed and elaborated
    in Barry Smith 1982 and Simons 1987. A detailed model of dependence is
    developed in Thomasson 1995.

11. States of affairs are "formal" structures in Husserl's ontology in *Logical Inves-
    tigations* (1900–1). They play a similar role in Wittgenstein's *Tractatus Logico-
    Philosophicus* (1921).

12. These formal categories are transformed, for example, in Whitehead's cate-
    gorial ontology, where "actual occasions" are tied in *nexus* to "forms" as well
    as to other actual occasions. See Whitehead 1978/1929. Whitehead's actual
    occasions replace Aristotle's substances, and his demi-Platonic forms tied to
    actual occasions do not reside in a Platonic heaven. Still, actual occasions
    would count as individuals in the present formal-material categories (al-
    though Whitehead posits his own system of categories with many differences
    from the simpler scheme sketched here). Thus, in the categorial ontology
    we have so far sketched, it remains to specify the status of species, qualities,
    relations, and so on. They might be treated as Platonic forms that can be
    exemplified by particular natural objects, or as Aristotelian universals that
    gain reality only insofar as they are instantiated, or as "tropes" or Husserlian
    "moments" (particularized universals) falling under the categories Species,
    Quality, and the like, or as Whiteheadian forms concretely tied into actual
    occasions. Alternatively, Platonized naturalism might posit Platonic forms as
    falling under the formal categories of Species, Quality, Relation, and so on,
    in an ontology of natural objects. Compare Linsky and Zalta 1995, with a
    specific formulation of a Mallian ontology that mixes abstract and concrete
    objects in a systematic ontology.

13. See the account of the world of common sense in Barry Smith 1995.

14. Compare my reconstruction of Husserl's ontology, where the same concrete
    individuals or events – in the world of facts – may instantiate the very dif-
    ferent ideal essences of Nature, Culture, Consciousness. See D. W. Smith
    1995.

15. See Kim 1993.

16. See the works by Husserl and Searle cited in the References. My approach
    (1989) defines a contextualist variant of internalism, which I now see as a
    third way, neither internalist nor externalist, emphasizing dependence or
    "grounding" of intentionality in context.

17. See Fodor 1994 and Dretske 1995.

18. The relations between the intentionality of consciousness and the intensionality of sentences about intentional states are detailed in Smith and McIntyre 1982.

19. I have analyzed these structures in detail (1989: chap. 2).

20. The reasoning in Quine's position is expressed in *Pursuit of Truth* (1992). See my appraisal 1994. Quine's thesis of the indeterminacy of translation was a further reason Quine (1960) gave for renouncing a science of intention, and thus phenomenology. I do not treat that problem here, although Quine's modified views (1991) lead, I think, to a different take on indeterminacy, notably where empathy plays a role in understanding others.

21. See the References for relevant works of the authors cited. Compare Fodor 1994: 14–15, summarizing some current attitudes.

22. See Quine 1969.

23. On the controversies surrounding connectionism, see Horgan and Tienson 1991.

24. Philosophy of mind in cognitive science has centered on the issue of how the mental might be reduced to the physical, focusing on what might be the structure of causal or computational process that defines mind. The structure of intentionality might then be *preserved* either as a language of thought ("sentences in the head," on Fodor's earlier view) or as a flow of physical "information" outside the head (Dretske's view, partly adopted recently by Fodor). Or intentionality might be eliminated (as the Churchlands have sometimes urged, assuming, mistakenly, that intentionality must have the structure of sentences in the head). But there is a wide range of options between these extremes, preserving some intentional structures and eliminating others. Moreover, the ontological issues are joined by questions about the relations among our different levels of *theory* about mind and brain. The complexity of these issues is brought out in Bickle 1992a, 1992b.

25. See Searle 1983, 1992.

26. See Quine and Ullian 1978 for the basic principles of framing knowledge in general and scientific theory in particular. In *Pursuit of Truth* (1992) Quine begins to make room for the intentional, but see my article (1994).

27. See my book (1989).

28. The new attitude is evident, for instance, in Crick 1994 and Churchland 1995.

29. See Flanagan 1992. The proclamation is a reviewer's remark printed on the back of the book. Flanagan's discussion is indeed sensitive to a wide variety of phenomenological issues about consciousness, and he explicitly says that "it is incredible to think that we could do without phenomenology altogether" (p. 12), proceeding to argue for how phenomenological claims should be combined with empirical results of neuroscience. However, many issues of phenomenological analysis of structures of consciousness are explored in great depth and detail in the writings of Husserl, Heidegger, Sartre, Merleau-Ponty and many others in the phenomenological tradition. Good science cannot pass over these results, any more than it can pass over significant empirical experiments such as those of split-brain patients or blind-sight.

30. See Chalmers 1995, announcing his forthcoming book (1996).

31. See D. W. Smith 1995.
32. Even if there are many metaphysically possible worlds, each is a unified system: that is the point of unionism. And even if the "many-worlds" version of quantum mechanics is true, the different parallel worlds are related in an ordered way, and the "world" we find ourselves in has its unity: that is the point here stressed by unionism. Unionism is thus orthogonal to the issue of other possible worlds in modal ontology (actualism denies that view) and to the issue of other parallel worlds of quantum mechanics. I would also argue that unionism is orthogonal to the issue of nominalism (often construed as ruling out abstract entities because they belong in a Platonic heaven apart from the natural world); I return to this point later but only briefly.
33. See Searle 1992.
34. As a skeptic might have it: see Rey 1988.
35. The intentionality of a thought depends on a background of not only personal beliefs and skills, but also attitudes and ideas extant in one's culture. This conception of background is explored in my "The Background" in Chapter 5. Compare Husserl's notion of horizon in *Ideas* I and II, reconstructed in Smith and McIntyre 1982, and Searle's notion of background (1983, 1992).
36. The term "meme" was coined by Richard Dawkins and championed by Dennett (1991).
37. A cognate ontology is that which underlies the PACIS computer system under development by Ontek Corporation. That ontology is the collective work of Charles W. Dement, Peter Simons, and myself. Parts of the ontology are previewed in Simons and Smith (with Charles W. Dement) 1993 and in Simons 1994–95, reflecting on a more recent version of our PACIS categories.

References

Aristotle. 1963. *Categories.* Translated by J. L. Akrill. Oxford: Oxford University Press.

Armstrong, David M. 1989. *Universals: An Opinionated Introduction.* Boulder: Westview Press.

1996. *A World of States of Affairs.* Cambridge: Cambridge University Press.

Armstrong, D. M., and Norman Malcolm. 1984. *Consciousness and Causality.* Oxford: Basil Blackwell.

Bickle, John. 1992a. "Revisionary Physicalism." *Biology and Philosophy* 7: 411–30.

1992b. "Mental Anomaly and the New Mind-Brain Reductionism." *Philosophy of Science* 59: 217–30.

Chalmers, David. 1995. "The Puzzle of Conscious Experience." *Scientific American* (December): 80–86.

1996. *The Conscious Mind.* Oxford: Oxford University Press.

Churchland, Patricia S. 1986. *Neurophilosophy: Toward a Unified Science of the Mind-Brain.* Cambridge, Mass.: MIT Press.

Churchland, Paul M. 1995. *The Engine of Reason, the Seat of the Soul.* Cambridge, Mass.: MIT Press.

Crick, Francis. 1994. *The Astonishing Hypothesis: The Scientific Search for the Soul.* New York: Scribner's and Sons.

Davidson, Donald. 1980. "Mental Events." In Davidson, *Essays on Actions and Events,* pp. 207–25. Oxford: Oxford University Press. First edition, 1970.

Dennett, Daniel. 1987. *The Intentional Stance.* Cambridge, Mass.: MIT Press.

1991. *Consciousness Explained.* New York: Little, Brown.

Dretske, Fred. 1981. *Knowledge and the Flow of Information.* Cambridge, Mass.: MIT Press.

1995. *Naturalizing the Mind.* Cambridge, Mass.: MIT Press.

Dreyfus, Hubert, ed. 1982. *Husserl, Intentionality and Cognitive Science.* Cambridge, Mass.: MIT Press.

Flanagan, Owen. 1992. *Consciousness Reconsidered.* Cambridge, Mass.: MIT Press.

Fodor, Jerry A. 1975. *The Language of Thought.* New York: Thomas Y. Crowell.

1994. *The Elm and the Expert: Mentalese and Its Semantics.* Cambridge, Mass.: MIT Press.

Horgan, Terence, and John Tienson, eds. 1991. *Connectionism and the Philosophy of Mind.* Dordrecht: Kluwer Academic Publishers.

Husserl, Edmund. 1963. *Ideas.* Translated by W. R. Boyce Gibson. New York: Collier Books. German original, 1913. Originally titled *Ideas Pertaining to a Pure Phenomenology and to a Phenomenological Philosophy, First Book.*

1970. *Logical Investigations.* Vols. 1 and 2. Translated by J. N. Findlay from the revised, second German edition. London: Routledge and Kegan Paul. New edition, edited with an introduction by Dermot Moran, and with a preface by Michael Dummett. London: Routledge, 2001. German original, 1900–1, revised 1913 (Prolegomena and Investigations I–V), 1920 (Investigation VI).

1989. *Ideas pertaining to a Pure Phenomenology and to a Phenomenological Philosophy, Second Book.* Translated by Richard Rojcewicz and André Schuwer. Dordrecht: Kluwer Academic Publishers. From the German original unpublished manuscript of 1912, revised 1915, 1928.

Kim, Jaegwon. 1993. *Supervenience and Mind.* Cambridge: Cambridge University Press.

Linsky, Bernard, and Edward Zalta. 1995. "Naturalized Platonism and Platonized Naturalism." *Journal of Philosophy* 0022-362X: 525–55.

McIntyre, Ronald. 1986. "Husserl and the Representational Theory of Mind." *Topoi* 5: 101–13.

Quine, W. V. 1960. *Word and Object.* Cambridge, Mass.: MIT Press.

1969. *Ontological Relativity and Other Essays.* New York: Columbia University Press.

1992. *Pursuit of Truth.* Cambridge, Mass.: Harvard University Press.

Quine, W. V., and Joseph S. Ullian. 1978. *The Web of Belief.* New York: Random House.

Rey, Georges. 1988. "A Question about Consciousness." In Herbert R. Otto and James A. Tuedio, eds., *Perspectives on Mind,* pp. 5–24. Dordrecht: D. Reidel.

Searle, John R. 1983. *Intentionality.* Cambridge: Cambridge University Press.

1992. *The Rediscovery of the Mind.* Cambridge, Mass.: MIT Press.

Simons, Peter. 1987. *Parts.* Oxford: Oxford University Press.

1994–95. "New Categories for Formal Ontology." *Grazer Philosophische Studien* 49: 77–99.

Simons, Peter, and David Woodruff Smith, with Charles W. Dement. 1993. "The Philosophical Foundations of PACIS." Paper presented at the International Ludwig Wittgenstein Symposium, Kirchberg am Wechsel, Austria, 19 August.

Smith, Barry, ed. 1982. *Parts and Moments.* Munich: Philosophia Verlag.

1995. "Common Sense." In Barry Smith and David Woodruff Smith, eds., *The Cambridge Companion to Husserl,* pp. 394–437. Cambridge: Cambridge University Press.

Smith, Barry, and David Woodruff Smith, eds. 1995. *The Cambridge Companion to Husserl.* Cambridge: Cambridge University Press.

Smith, David Woodruff. 1989. *The Circle of Acquaintance.* Dordrecht: Kluwer Academic Publishers.

1994. "How to Husserl a Quine – and a Heidegger Too." *Synthese* 98 (1): 153–73.

1995. "Mind and Body." In Barry Smith and David Woodruff Smith, eds., *The Cambridge Companion to Husserl,* pp. 323–73. Cambridge: Cambridge University Press.

Smith, David Woodruff, and Ronald McIntyre. 1982. *Husserl and Intentionality.* Dordrecht: D. Reidel.

Thomasson, Amie L. 1995. "The Ontology of Fiction: A Study of Dependent Objects." Ph.D. dissertation, University of California, Irvine.

Whitehead, Alfred North. 1978. *Process and Reality.* Edited by David Ray Griffin and Donald W. Sherburne. New York: Free Press, Macmillan. First edition, 1929.

7

Consciousness and Actuality

Abstract. Alfred North Whitehead elaborated an ontology of process that we can use to frame a novel ontology of consciousness and cognition. Whitehead amplified the familiar notion of temporal process ("all is flux," echoing Heraclitus) in an ontology that presages today's dynamical systems theory, with obvious application to both events in nature and events of consciousness. But Whitehead also developed a different and deeper notion of process: what I call "ontological becoming," as opposed to temporal becoming. Ontological becoming, I propose, is the formal structure realized in the material structure of temporal becoming. Both apply to processes in general and to processes of consciousness in particular. From Whitehead's ontology we can thereby begin to learn how we might develop an ontological "systematics," a broad and fundamental framework that gives consciousness its due along with the rest of the world.

Segue. In "Intentionality Naturalized?" we argued that the theory of mind needs a broad and sensitive ontology that places consciousness in the structure of the natural world while preserving the phenomenological character of conscious experience. As we saw in that essay, one approach to a sufficiently rich ontology borrows from Husserl, developing not only his analysis of the structure of intentionality, but also his distinction between formal and material ontology (see "Basic Categories" in Chapter 8). A still more radical approach is found in Whitehead's process ontology, to which we turn here, looking for a model of how to analyze consciousness

I wish to acknowledge my debt to the late John D. Goheen of Stanford University. I first studied Whitehead's metaphysics in a course taught by Professor Goheen, and the image of "prehension" has stayed with me for three decades. In recent years the image has emerged, with a life of its own, in the very practical needs of the PACIS database project of Ontek Corporation. I thank Charles W. Dement of Ontek for many discussions of Whitehead's ontology and its role in prefiguring the ontology of PACIS. My thanks finally to Liliana Albertazzi for inviting me to write this essay, in a sense for seeing its need.

in a wider structure of the world. In the long run the aim is to borrow from Whitehead's "depth" ontology (as I shall call it) to build a framework for an ontology that respects both physical and phenomenological structure.

Cognition in the Structure of Existence

If cognitive science is the science of cognition, what is cognition? Whitehead's answer is drawn directly from his fundamental ontology: cognition is a special form of what he calls "prehension," taking more particular forms in conscious perception, thought, imagination and other psychological processes.

From Whitehead's sweeping metaphysics I propose to appropriate key elements and reconstruct from these a fundamental ontology that applies to consciousness in cognitive processes. This Whiteheadian ontology has shaped a novel approach to fundamental ontological categories used in database design or knowledge-based computer systems.[1] And the Whiteheadian approach to conscious processes is instructive for the dynamical systems approach in cognitive science, which has been proposed as an alternative to both the neural-network approach and the language-of-thought approach to modeling cognitive processes.[2]

Whitehead's main concern was not cognitive process, but process in general. Within a fundamental ontology of process, he turned to processes of consciousness, perception, belief, and emotion – almost in passing as he sought to limn the truly ultimate structure of reality. For those raised in empirical science, or in philosophy oriented to science and modern logic, Whitehead's ontology will seem the most extreme of a priori speculative metaphysics, written moreover in a philosophical prose that is as complex as Russell's is simple. Still, the mathematical wizardry in *Principia Mathematica* was more Whitehead than Russell, and the purity of Whitehead's ontological vision (shaped by his mathematician's brain) is kindred in soul with the physical vision of an Einstein or a Hawking. It is this pure ontology that I want to frame, in order to construct from it, along lines Whitehead prescribed, an ontology of consciousness in various forms of cognition.

The resulting Whiteheadian ontology does not naturalize consciousness by reducing it to physical process. Rather, it places all of nature *cum* consciousness in a world formed by one ontological process of "becoming" through "prehension."

Whitehead's Background and Context

Alfred North Whitehead (1861–1947) is known as a principal founder of mathematical logic, through the monumental *Principia Mathematica* (1910–13), which he coauthored with the younger Bertrand Russell. Whitehead and Russell exerted an indirect influence through logic on computer science and thus on the computer model of mind that informs cognitive science today. Whitehead's later years, however, saw him engaged in abstract metaphysics, in work that is usually treated quite separately from his early work in mathematics and then mathematical logic. Here I address Whitehead's ontology as developed in *Process and Reality* (1929; delivered as a lecture series in 1927–28).

Whitehead posited a basic structure of "process" applied to all reality. As a mathematician he was keenly attuned to the physics of relativity and quantum mechanics emerging in his day. As a British-born philosopher he had his eye also on the early modern philosophers, from Descartes to Locke and Hume, for whom mental activity was salient. Yet as a philosophical cosmologist he looked to the ancient Greeks, to Plato and the pre-Socratics, especially Heraclitus, who famously proclaimed that everything is flux. Reflecting a further sensibility of the nineteenth century, Whitehead called his metaphysical-cosmological system "the philosophy of organism," and he did indeed think of the cosmos itself as alive in a fundamental sense. Today his system is often called "process philosophy" and is influential in theology, because Whitehead defined a place for God in the cosmos, as the first being formed by the process of "becoming."

With an eye to cognitive science, as opposed to metaphysical system, consider from Whitehead's era three very different approaches to the exploration of the mind. Wundt's experimental psychology used introspection to study sensations. Freud's psychoanalysis used hypotheses about drives and infantile experience to explain adult neuroses. And Husserl's phenomenology used the technique of reflection on one's own experience to analyze the structure of consciousness, including the dynamics of time consciousness. Here were three starkly different "sciences" of the mind, differing in both aim and method. Yet each had to address the dynamics of changing and interacting states of mind, and today's cognitive science carries the analysis of mind into mathematical-computational models of the dynamics of cognitive process. Whitehead's ontology of process ought to shed light on the most basic structure of mental process as a flow of cognitive occasions.

Whitehead drew where possible on classical ideas: in Plato, Aristotle, Heraclitus, Descartes, Leibniz, Locke, Hume. Yet the proximate influence on his conception of process was the French philosopher Henri Bergson. In *Matter and Memory* (1896), Bergson painted a metaphysics featuring "durations" rather than classical enduring natural objects, what Aristotle called "substances" (*ousia*). The world is a vortex of passing events in interaction with one another, each bearing within it "images" of others. Bergson envisioned a monism whereby this metaphysical picture applies equally to physical events and to mental events. Physical events were soon to be described in relativity theory and then in quantum physics, while mental events were soon to be described in phenomenology, in psychoanalysis, and ultimately in what has recently emerged as cognitive science. Bergson would have wished his metaphysics to apply to the dynamics of quantum and relativistic physical phenomena and equally to the dynamics of perception, imagination, thought, and desire. Bergson expressed his vision in vivid impressionistic prose, mirroring in metaphysical language the flow of painting in French impressionism and fauvism. I find Whitehead's metaphysical prose impressionistic yet mathematical. Whitehead's technical terms are often vividly, disconcertingly, mentalistic. Yet his terminology reflects a Bergsonian style of monism rather than the varieties of absolute idealism that ran from Hegel to F. H. Bradley and J. M. E. McTaggart in England. Like absolute idealism, Whitehead's philosophy of "organism" preached the radical interaction of all things; however, for Whitehead, the ultimately actual entities are neither mental nor physical, but constitute what we know as the mental and the physical alike.

McTaggart taught philosophy at Cambridge when Whitehead taught mathematics at Cambridge. Like McTaggart, as well as Bergson, Whitehead focused on the process of coming to pass in time, where the truly "actual" – what deserves the name – is in process, becoming rather than already become. McTaggart (1927) argued that only the present is actual, that past and future are unreal, indeed that time itself is unreal because a time or occurrence cannot be first future, then present, then past, as these temporal attributions are mutually contradictory. Whitehead (1927–28) held, by contrast, that past and future are perfectly real but not "actual," or actualizing, that is, in process of becoming. Still, Whitehead had his eye on bigger fish when he framed actuality as *becoming*: his concern was not only coming to pass in time, but coming to be an entity at all, of which coming to be an entity in temporal flux is a special case.

Whitehead's fundamental ontology of process, I want to suggest, provides formal tools that can be used to analyze the substantive claims about process, including processes of consciousness, put forth in empirical or experiential studies. By "formal," however, I mean pertaining to *formal ontology*, as opposed to formal languages of logic, mathematics, or computer science. Here I draw on Husserl's distinction between formal and material ontology. "Formal" ontology seeks invariant structures that are realized in various domains of entities, whereas "material" ontologies analyze ontological structures within a given domain (such as mind, body, culture, or what have you).[3] Whitehead's account of the process of becoming an entity is, I shall try to show, a formal ontology of "process," and consciousness inherits this structure of process.

The Outlines of Whitehead's Ontology

Whitehead's ontology in *Process and Reality* is framed by an elaborate system of *categories* (1978: 18–30). I can address only that part of the system which I appropriate for present purposes.

The most fundamental category, dubbed "the Category of the Ultimate," is that of Creativity or Becoming. Becoming is the process by which *many* entities become – come together to form – *one* actual entity. Becoming is also called concrescence (p. 21). Importantly, the category of becoming "replaces Aristotle's category of 'primary substance'" (p. 21). Aristotle's paradigm of a primary substance (*ousia*, or being) is a concrete being such as an individual human organism; such a being is a member of a species, in this case humanity, and species themselves belong to the category Aristotle called secondary substance. For Aristotle, a primary substance is something that endures through time and may change its qualities or "accidents" such as its relations to other things or its location; and it is a fusion of matter and form, the matter undergoing change in form. Now, Whitehead's leading claim is that such beings are not the most fundamental type of thing in the world. More fundamental are the processes of *becoming* that form all actual entities. Becomings, not "substances," are the ultimate constituents of the world, the cosmos. Aristotelian substances are, for Whitehead, the product of many processes of becoming, and this view is an abstract version of modern physical theory, for which everyday objects are products of relativistic, quantum phenomena of a more fundamental and very different kind.

Presupposing the category of Becoming are the Categories of Existence (p. 22). There are eight such categories, of which I shall focus on three: Actual Entity, Eternal Object, and Prehension (p. 1). An actual entity, or actual occasion, is a concrete entity rather like a "point" event (p. 2). An eternal object, or form, is what Plato called a form (*eidos*), but Whitehead conceives a form, or "form of definiteness," as a "pure potentiality" for being "determined" in a specific way. For Plato, a form is an actual but eternal entity (residing in a Platonic heaven, as Plato is often parodied, a parody Whitehead rejects). For Whitehead, however, a form is not an actual albeit eternal entity, but rather a pure potential for actual entities: the way an actual entity could be – which is "eternal" insofar as it is not realized in an actual entity (p. 3). Now, prehension is the tie that binds everything together; it is the way in which things are, most fundamentally, related. If you will, prehension is simply the fundamental form of relatedness, and particular prehensions are the basic "concrete facts of relatedness." Thus, actual entities prehend other (prior) actual entities, and they prehend forms as well. (More on all this shortly.)

Presupposing both the Category of Becoming and the Categories of Existence are two further types of categories. The Categories of Explanation are twenty-seven axiomatic conditions, or types of interaction, among actual entities, eternal forms, prehensions, and the like – basic conditions laid down by axioms in the ontology. And the Categoreal Obligations are nine preconditions laid down, as it were, by further axioms. These categories together weave the story of reality as told in Whitehead's ontology. Importantly, Whitehead *ranks* different types of categories: most basic is the Category of Becoming, next are the Categories of Existence, then the Categories of Condition and then those of Precondition (as I am calling them).

With this ranking of categories, Whitehead moves well beyond Aristotle, who launched the idea of ontological categories with his famous list:

Primary Substance (Individual), Secondary Substance (Species), Quantity, Quality, Relative, Where, When, Position, Having, Acting, Being-Acted-Upon.

Aristotle's term *kategoriai* originally meant simply "predicates," or types of entities. By contrast, what are called categories in biology today – Species, Genus, Family, Order – are ranks of taxa (groups of living things), defining their place in the classification scheme. Indeed, Whitehead's categorial scheme shows something of the complexity that has since arisen in modern biological systematics, or higher classification. The great biologist Ernst Mayr even refers to Whitehead while distinguishing the evolutionary roles of species as opposed to higher taxa.[4] Where systematics in biology is the science of diversity of living beings, we may conceive *ontological systematics* as the science of diversity of all beings, classifying things of all types and categories. A ranking of categories is then part of the task of ontological systematics, and of "formal" ontology in the sense that Husserl had in mind.

Assuming Whitehead's scheme of categories, let us turn to his account of actual entities. To be "actual" is to be active, to be in process, to be part of the flux that constitutes all actuality. As Heraclitus insisted, all is flux, "all things flow" (p. 208). Thus, in Whitehead's vision, the atoms of the universe are not particles but points in the flux, points of flow: processes of becoming, or actual entities or occasions in the process of becoming, coming into actuality. Points along the flow are bound together by prehension: later occasions or actual(izing) entities *prehend* or "feel" earlier ones. We think of a physical "feeling" in gravitation or electromagnetic attraction, but prehension is – I should like to say in interpretation – the *formal* connection that is realized as gravitational or electromagnetic connection, or physical causation, in the relevant scenario of physics. Again, we think of an emotional "feeling" in personal attraction, where prehension is the formal connection that is realized as a bond of friendship, in the relevant scenario of psychology. And we may think of a cognitive "feeling" in conscious perception of an object, where prehension is the formal connection that is realized as a bond of *intentionality* – to use Husserl's term, which has now gained currency in phenomenology, philosophy of mind, and cognitive science, but was unknown to Whitehead. The root meaning of "intention" is to aim or reach toward something; this is exactly the metaphor behind Whitehead's term "prehension," as he initially defines a prehension as having "a vector character" (p. 19). Whitehead looks to Leibniz's *Monadology* (p. 19) as a model: indeed, his ontology is like Leibniz's but with the monads' windows open. However, whereas Leibniz looks to mental operations for a model of the physical, producing his unique brand of idealism, Whitehead sees the notion of prehension as striking a "balance" between mental and physical operations (p. 19). A better way of putting the point, I propose, is to say that prehension is the common formal structure that is realized in operations of both mental and physical "feeling."

Now, Whitehead recognizes different types of prehension (p. 1). There are positive prehensions, called "feelings" (illustrated provisionally here). But there are also negative prehensions, which block or "eliminate" things from feeling. We address only positive prehensions, or feelings (p. 2). An actual entity E prehends not only other actual entities but also eternal objects or forms. Prehending an actual entity is called "physical" prehension, whereas prehending a form is called "conceptual" prehension (as discussed later). E's prehending a form F is Whitehead's variation on what we call, since Plato, E's exemplifying F. In Whitehead's idiom, E "feels" the form F; alternatively, F "ingresses" into E.

For Plato, forms exist eternally and thus in a different realm than particulars; for Aristotle, *contra* Plato, forms (species, accidents, locations) exist in the concrete world but only insofar as instantiated by particular substances. But Whitehead places both concrete actual entities and abstract forms precisely in the flux that is reality. The difference is that a form is a *potential* for definiteness in an actual entity, whereas an actual entity is a definite realized entity, that is, in process of becoming and thereby being fully "determined." Later we address, in Whitehead's fine-grained analysis, a hybrid entity that combines the actual and the potential. (I leave the details as a surprise to the reader.)

The prior actual entities prehended by a given actual entity E form the "datum" of E, or the "actual world" of E. The constitution of E – its "real internal constitution" (p. 25) – includes the pattern of prehensions in which E feels the actual entities in its actual world and also feels the forms that are realized in it. Importantly, there is an ontological relativity in Whitehead's scheme: the actual world of E is not the actual world *tout court*, for every actual entity has its own "actual world." The point to remember is that Whitehead's ontology is "tensed," or rather indexicalized. Recall McTaggart's distinction between two structures of time, or two types of temporality: times are ordered in what he called the A-series, where times are past, present, or future; and times are ordered in the B-series, where one time is before another time. It is customary to say that B-series time is time viewed from outside the space-time framework, as it were eternally, whereas A-series time is time viewed from within the space-time framework, and thus indexically. Now, actual entities are ordered by becoming. To add a dose of McTaggart to Whitehead, actual entities are "present," or presently actualizing. This currency defines "actuality" for Whitehead. (That is in fact one of the meanings of "actual" – in French and German, current events.) Think, then, of flux as a transition from past (formerly) actual entities into present (currently) actual entities and toward future (potentially) actual entities, always informed by the realization-in-progress of potentials called forms.

There is, however, a second type of flux or "fluency," which we consider shortly: not the *transition* from past actualities to a presently unfolding actuality but the *concrescence* wherein many actual entities come together to form one actual entity (p. 210).

Whitehead's ontology seems apt for the world of twentieth-century physics, for points of matter-energy in a field of space-time in a relativistic quantum framework. But how does it apply to the everyday world of objects and properties, the world described by Aristotle's metaphysics, which is a more theoretical version of our commonsense ontology? Whitehead did not simply throw out Aristotelian metaphysics. He said Aristotle's category of substance is not fundamental, but he wanted to preserve everyday objects or "substances" as emergent from more fundamental "actual entities." Moreover, he preserved the Platonic category of form. For Whitehead, actual entities have forms, as did Aristotelian substances, but forms are potentials for actual entities.

Ontological versus Temporal "Becoming"

All things are in flux, "all things flow," Whitehead proclaims (1978: 208), echoing Heraclitus. Yet Whitehead immediately introduces a crucial, and I think profound, distinction:

> [T]here are two kinds of fluency. One kind is the *concrescence* which, in Locke's language, is "the real internal constitution of a particular existent." The other kind is the *transition* from particular existent to particular existent. This transition, again in Locke's language, is the "perpetually perishing" which is one aspect of the notion of time, and in another aspect the transition is the origination of the present in conformity with the "power" of the past. (p. 210)

Thus, Whitehead distinguishes two fundamentally different types of "flux" in the process of "becoming." We might say that transition from one entity at one time to another entity at a later time defines *temporal becoming*, as the later entity comes into being in time. And we might say that concrescence of one entity out of others defines *ontological becoming*, as the one comes into being on the basis of the others.

Suppose we treat the process of "flowing" as a *function* or *operation* in the mathematical sense (Whitehead being a mathematician, after all). Then *temporal becoming* is an operation that assigns to one set of occasions $\{O_1, \ldots, O_n\}$ another occasion O^*. The occasions O_i in the original set, which are the arguments of the function, are "prior" in time, "past" occasions relative to O^*, which is "present" or "actual (now)." By contrast, *ontological becoming*, or concrescence, is an operation that assigns to one set of actual entities $\{E_1, \ldots, E_n\}$ another actual entity E^*. The entities E_i compose the "datum" or "actual world" of E^*. Notice that the term "occasion" carries temporal connotations, whereas "entity" does not.

Whitehead assumes "the ontological principle" (pp. 19, 24) that every actual entity has its "reasons" or grounds, the entities out of which it is formed by concrescence: "no actual entity, then no reason" (p. 19). Clearly, this principle constrains ontological rather than temporal becoming.

Although Whitehead's language of "process" resonates with the sound of transition through time, his ontology actually treats concrescence, or ontological becoming, as what is fundamental. Indeed, extension in time and space is itself something that must come into being, or become, and its becoming is not temporal. That is, temporality becomes, but its becoming is not temporal. In Whitehead's words:

[I]n every act of becoming there is the becoming of something with temporal extension; but... the act itself is not extensive, in the sense that it is divisible into earlier and later acts of becoming which correspond to the extensive divisibility of what has become. (p. 69)

This extensive continuum [of time] is one relational complex in which all potential objectifications find their niche. It underlies the whole world, past, present, and future.... It is not a fact prior to the world; it is the first determination of order – that is, of real potentiality – arising out of the general character of the world. In its full generality beyond the present epoch, it does not involve shapes, dimensions, or measurability; these are additional determinations of real potentiality arising from our cosmic epoch. (p. 66)

[T]he extensive continuity of the physical universe has usually been construed to mean that there is a continuity of becoming. But if we admit that "something becomes," it is easy, by employing Zeno's method, to prove that there can be no continuity of becoming. There is a becoming of continuity, but no continuity of becoming.... In other words, extensiveness becomes, but "becoming" is not itself extensive. (p. 35)

Thus, extension in time is a fully "general" structure of the world. And that structure of temporality must itself *become*. It is in fact "the first determination of order": first in ontological priority. Time is not, however, the first thing to exist in time, first in temporal priority.

We can clarify Whitehead's distinction of two kinds of flux if we distinguish between formal and material ontology. Temporal becoming is, in these terms, a "material" process, and ontological becoming is a "formal" process: concrescence, or ontological becoming, is the formal structure that applies to all actual entities, in particular to the material domain of entities in temporal becoming. Indeed, for Whitehead, concrescence is the most fundamental *formal* ontological structure of the world: the category of "the ultimate."

While the process of ontological becoming is the heart of Whitehead's ontology, the most salient application of the ontology is to temporal becoming, or "process" in the familiar sense, of which mental process is a special case. This dual sense of "process" plagues the text of *Process and Reality*. May I suggest a resolution of this interpretative quandary? If we analyze the two types of flux in the manner outlined here, then the formal structure of ontological becoming is reproduced at a different level in the material structure of temporal becoming. There is a recursion of basic structure in this Whiteheadian analysis of "becoming": not only is temporal becoming the product of ontological becoming (time becomes, takes ontological form), but its *form* is the same as that of ontological becoming, or concrescence. However, even if this form does recur at both levels (and I am not sure it really does), it remains misleading to use the same term "process" for both the becoming of a temporal transition and the transitioning itself.

I share Whitehead's appreciation of the pre-Socratic philosophers of ancient Greece, whose concerns with truly fundamental issues like "the one and the many" have been eclipsed in more recent philosophy. But where Whitehead nods to Heraclitus on "flux," I bow to Anaximander of Miletus on "origin," or ontological basis. In Anaximander's metaphysical vision, all types of beings in the world (including the most basic physical-chemical elements as he knew them) have their origin, or *arche*, in something he called the *apeiron*, the "nonlimited" or "nondetermined." In this vision I find roots of the notion of ontological becoming, whereby every being or entity (*ousia* in Aristotle) is based in or arises from (what I like to call) primordial *modes of being*, which are not entities at all but rather ways in which entities become entities, and so are the bases of beings. I believe Whitehead's "category of the ultimate" – Becoming – follows this Anaximandrian architecture. Thus, "determinate" actual entities, extended in time and space-time, are constituted in ways of prehending that are not themselves "determinate" entities.[5]

Prehension or "Feeling"

The ontological structure of an actual entity or occasion, for Whitehead, is that of a complex defined by relations of prehension. Indeed, an actual entity or occasion is a "cell," in a sense akin to the biological. As Whitehead puts it:

[T]he cell is exhibited as appropriating for the foundation of its own existence, the various elements of the universe out of which it arises. Each process of appropriation of a particular element is termed a prehension. The ultimate elements of the universe, thus appropriated, are the already constituted actual entities, and the eternal objects. The process itself is the constitution of the actual entity; in Locke's phrase, it is the "real internal constitution" of the actual entity. (1978: 219)

An actual entity is in this way "constituted" by its pattern of prehensions of other actual entities, and also eternal objects or forms. Thereby it *becomes* "one" actual entity, formed by concrescence out of "many" actual entities and forms.

Prehension – called "feeling" or "experience" (if positive) – is not per se a psychological phenomenon. Prehending something is not the same as being *conscious of* something, what Husserl dubbed intentionality. As we shall see, consciousness is a special form of prehension, and not vice versa.

There are two primary species of prehension or feeling: physical and "conceptual." Physical prehension is what we think of as causation, whereas conceptual prehension is what we think of as being informed by a property or form. I say "what we think of" because Whitehead wants to underwrite our familiar, Aristotelian metaphysics by a very different ontology of prehension. In Whitehead's words:

A simply physical feeling is an act of causation. The actual entity which is the initial datum is the "cause," and the simply physical feeling is the "effect," and the subject entertaining the simply physical feeling is the actual entity "conditioned" by the effect. This "conditioned" actual entity will also be called the "effect." (p. 236)

A conceptual feeling is feeling an eternal object in the primary metaphysical character of being an "object," that is to say, feeling its capacity for being a realised determinant of process. (p. 239)

Whitehead thinks of a Platonic form or "eternal object" – what we call a kind, quality, or relation – as a *potential* or *capacity* for being determined thus and so (p. 23), and this is what is felt "conceptually." Rather than residing in a Platonic heaven, it exists here in the one Heraclitean flux that is the world. Its prehension Whitehead calls "conceptual"; "formal" or "eidetic" would avoid the suggestion of a mental activity of conceptualization, but Whitehead's terminology deliberately flouts the distinction between the mental and the nonmental.

Now, every actual entity is formed by concrescence through prehensions of *both* actual entities and forms, as depicted in Figure 7.1. Note the

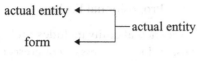

FIGURE 7.1. Concrescence.

mixing of mentalist and physicalist terminology as Whitehead makes this point:

> In each concrescence there is a twofold aspect of the creative urge. In one aspect there is the origination of simple causal feelings; and in the other aspect there is the origination of conceptual feelings. These contrasted aspects will be called the physical and mental poles of an actual entity. No actual entity is devoid of either pole; though their relative importance differs in different actual entities. Also conceptual feelings do not necessarily involve consciousness; though there can be no conscious feelings which do not involve conceptual feelings as elements in the synthesis. (p. 239)

So every actual entity "feels" both physically and conceptually, feeling both (other) actual entities and eternal forms. Feeling a form is called "mental" but does not involve consciousness, or indeed what we think of as mental activity in thought or perception. Here, in the most elementary structure of the world, Whitehead nails down the monism we noted earlier – under the influence of Bergson rather than Spinoza (who assumed one *substance*). Furthermore, we begin to see the niche Whitehead has in mind for consciousness: whatever is special about consciousness, it consists in "conscious feelings," and these always involve feeling forms.

Furthermore, all the world – from actual entities on up – is defined by prehensions of both forms and actual entities: "Conceptual feelings and simple causal feelings constitute the two main species of 'primary' feelings. All other feelings of whatever complexity arise out of a process of integration which starts with a phase of these primary feelings" (p. 239).

Whitehead holds a kind of atomism (pp. 19, 35, 238), because actual entities are atomic and everything complex is built up from prehensions of actual entities and forms. But the atomism is not that of physics, where elementary particles are bound together by physical-chemical bonds (physical feelings of causation). Nor is the atomism that of logical atomism, as propounded by Wittgenstein in the *Tractatus* (1921) or Russell in "The Philosophy of Logical Atomism" (1918), where objects and properties or relations are bound together by instantiation and the resulting "states of affairs" or "facts" are then further bound together by logical operations such as "and."

Propositional Feeling

The constitution of every actual entity includes prehensions of both actual entities and forms, and the process of concrescence constitutes the
entity through these prehensions. Now, concrescence itself takes place
in three phases that Whitehead calls reception, supplementation, and
satisfaction (1978: 211–15). The concrescence consists in "reception" of
a datum comprising other actual entities and forms, "supplementation"
of the reception by something like interpretation, and completion or
"satisfaction" producing the new actual entity. (These are not temporal
but logical or rather ontological "phases" of the process.) Concrescence
is not a psychological operation, yet Whitehead analyzes concrescence
into a structure like that of cognition, with a distinctly Kantian twist, and
it is in one phase of concrescence that consciousness itself may emerge in
some cases. Let us see how.

The supplementation phase of concrescence Whitehead analyzes into
two subphases called "aesthetic" and "intellectual" (pp. 213–14). Aesthetic supplementation "has welded the data into a new fact of blind
feeling," and then, "This phase requires an influx of conceptual feelings
and their integration with the pure physical feelings." This integration of
physical and conceptual feelings is "intellectual sight." This integrative
higher-phase feeling Whitehead calls "propositional feeling" (p. 214).

Specifically, a form or eternal object is a capacity for "determination,"
that is, being propertied or related in a specific way. When a form is realized in an actual entity, that entity is determinate in respect of that form.
And a *propositional* feeling is a higher-phase prehension of the propositional integration of a form in a determinate actual entity: "An eternal
object realised in respect to its pure potentiality as related to *determinate*
logical subjects is termed a 'propositional feeling' in the mentality of the
actual occasion in question" (p. 214).

In the web Whitehead weaves, we discern the picture in Figure 7.2.
Thus, propositional feeling is a higher-phase prehension that consists in
feeling, with intellectual sight, the integration of the received actual entities and forms. That integration is called a "proposition" (p. 256) and the
prehension of a proposition is a propositional feeling. Figure 7.3 depicts
this form of prehension, whose fine structure is depicted in Figure 7.2.
(We return to Whitehead's ontology of propositions later.)

In this scheme for propositional feeling we hear the echo of Kant's
famous dictum, "Intuition without concepts is blind; concepts without
intuition are empty." Kant's concern was the structure of cognition that

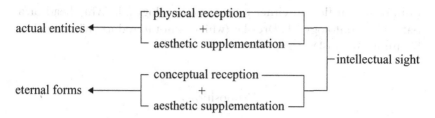

FIGURE 7.2. Propositional feeling.

(form + actual entity) = proposition ◄——————— (propositional feeling)

FIGURE 7.3. A proposition and its prehension.

synthesizes sensations and concepts, or sensibility and understanding. This structure applies to objects in nature precisely because, according to Kant's transcendental idealism, natural objects are phenomenal projections of cognition. But Whitehead inverts this order of priority: cognitions carry this structure precisely because all actual occasions have the structure of concrescence, which with intellectual sight begins to constitute an occasion of cognition or consciousness. (See his note on p. 215.) According to Whitehead, it is in this niche of concrescence, formally defined, that the structure characteristic of consciousness begins to emerge.

If Whitehead's scheme seems to force a mentalistic structure onto concresence itself (conceptual feeling is "mental," propositional feeling is part "intellectual"), consider a recent program in the philosophy of cognitive science. The Shannon-Weaver notion of "information" (roughly a conditional probability of this given that) is the basis of Fred Dretske's analysis of knowledge as a "flow of information."[6] In Dretske's ontological vision, cognition is a complex transfer of "information" from the environment to the knower, but this information-processing is conceived as a "natural" process, like causation, and the "information" transferred is not what we usually think of as a "proposition" in a mind. Well, in Whitehead's system, prehension itself is a kind of flow of information, as the datum of actual entities (and forms) that are prehended by an actual entity is the "reason" or foundation of the entity (the condition of this entity given those). Thus, we could see the *formal* structure of the process of information flow in either a monistic ontology à la Whitehead or a naturalistic ontology à la Dretske (which is not to reduce Whitehead to Shannon-Weaver-Dretske).

Propositions

The "objective datum" of a propositional feeling – if you will, the object of a propositional feeling – is what Whitehead calls a *proposition* (1978: 256). Specifically:

> An "impure" prehension arises from the integration of a "pure" conceptual prehension with a physical prehension originating in the physical pole. The datum of a pure conceptual prehension is an eternal object; the datum of an impure prehension is a proposition, otherwise termed a "theory." (p. 184)

> A proposition is a new kind of entity. It is a hybrid between pure potentialities [or forms] and actualities. The definite set of actual entities involved are called the "logical subjects of the proposition"; and the definite set of eternal objects involved are called the "predicates of the proposition." The predicates define a potentiality of relatedness for the subjects. The predicates form one complex eternal object: this is "the complex predicate." The "singular" proposition is the potentiality of this complex predicate finding realisation in the nexus of reactions between the logical subjects, with assigned stations in the pattern for the various logical subjects. (pp. 185–86)

Here are echoes of the voice(s) of *Principia Mathematica.*

But Whitehead's later ontology gives "propositions" an utterly different role than does traditional logic: "Indeed Bradley does not mention 'propositions' in his *Logic*. He writes only of 'judgements.' Other authors define propositions as a component in judgement. The doctrine here laid down is that, in the realisation of propositions, 'judgement' is a very rare component, and so is 'consciousness'" (p. 184). Thus, Frege posited a realm of "thoughts" that serve as "sense" (*Sinn*) expressed by language but are entities distinct from both the physical and the mental. Again, Husserl posited a realm of "noemata" that serve as "sense" (*Sinn*) or content of mental acts including judgments and are expressible in language but are distinct in ontological kind from mental acts, physical objects, and cultural institutions (including words). But Whitehead's conception of propositions is strikingly different, and genuinely novel, because a proposition per se is neither component nor content nor object of a mental act of judgment or consciousness or of a speech act or linguistic expression. Even in causal "feeling," a proposition is "what is proposed" for feeling.

Whitehead insists:

> The term "proposition" suits these hybrid entities, provided that we substitute the broad notion of "feeling" for the narrower notions of "judgement" and "belief." A proposition is an element in the objective lure *proposed for feeling*, and when admitted into feeling it constitutes *what is felt*. . . . Judgement is the decision admitting a proposition into intellectual belief. (p. 187)

We might call this hybrid entity a *state of affairs*, following a well-established tradition in Austrian philosophy, including Husserl and the early Wittgenstein, for whom a state of affairs (*Sachverhalt*: literally "things related") is a complex entity formed by a number of objects instantiating a determinate relation (or one object instantiating a determinate property or species). However, Whitehead joins Russell in importing this notion into philosophical English under the term "proposition." Unfortunately, we say in English that a proposition is "true" but a state of affairs "obtains" or is "actual." Nonetheless, Whitehead presses his terminology into the virgin territory of prehension loosed from thought or consciousness:

> The interest in logic, dominating overintellectualised philosophers, has obscured the main function of propositions in the nature of things. They are not primarily for belief, but for feeling at the physical level of unconsciousness. They constitute a source for the origination of feeling which is not tied down to mere datum. A proposition is "realised" by a member of its locus, when it is admitted into feeling.

> There are two types of relationship between a proposition and the actual world [or datum] of a member of its locus. The proposition may be conformal or non-conformal to the actual world, true or false. (p. 186)

Thus, Whitehead finds for "truth" a distinct role from that of "actuality" (becoming, concrescence), even while defining a proposition as rather like a state of affairs. Remember, the "actual world" of an actual entity is the nexus of (other) actual entities prehended by the entity. So a proposition felt by one actual entity is *true* just in case it is conformal to the relevant actual world of the entities that are logical subjects of the proposition felt.

Here Whitehead divines a subtle line between truth and actuality. It is natural to think, in a classical vein, that a proposition is a content of thought or language that *represents* a state of affairs beyond thought or language. Then a proposition is true if and only if it conforms to the actual state of affairs it represents. If there is no actual state of affairs to which a proposition corresponds, then the proposition is false. On a modal variation, a false proposition may yet represent a possible but nonactual state of affairs. However, on Whitehead's approach, a proposition is a hybrid of

actual entities and potentialities, that is, forms as potentialities for determination as thus and so in actuality. And this hybrid stands ontologically,
modally, between achieved actuality and mere potentiality. A proposition,
on this view, is neither a possible state of affairs nor a "proposed" state of
affairs. A possible state of affairs lacks the modal character of potentiality,
or *tending toward* actuality; and a proposed state of affairs has still a different modal character, if understood as an object *intentionally contained in*
a mental act of judgment or supposition or imagination, or intentionally
projected from and so dependent on a propositional act of consciousness. No, for Whitehead, a proposition is "proposed for feeling" in the
sense that it is a not an actual state of affairs but is poised in readiness
for being realized or formed through the process of conscrescence into
a nexus of actual entities in determinate relation. And this status is ontologically prior to any conscious act of judgment, imagination, or the
like.

No stranger to logic, Whitehead recognizes quantified as well as singular propositions: "A 'general' proposition only differs from a 'singular'
proposition by the generalisation of 'one definite set of actual entities'
into 'any set belonging to a certain sort of sets.' If the sort of sets includes
all sets with potentiality for that nexus of reactions, the proposition is
called 'universal'" (p. 186). So Whitehead recognizes not only singular
propositions (involving definite actual entities as "subjects") but also uni-
versally quantified propositions (involving "any" definite actual entities)
and also existentially quantified propositions (involving "some set of actual entities" but not a definite set) (see pp. 257–58). Also, Whitehead
recognizes "negative" as well as "positive" prehensions (p. 23) and thence
affirmative and negative perceptions and judgments in feeling affirmative
and negative propositions (pp. 161, 243, 270–74). So, with truth and the
familiar logical forms of propositions, Whitehead articulates the structures of modern logic: all within the realm of propositions defined as
ontological structures prior to judgment or thought content or indeed
linguistic expression!

If states of affairs are just "out there," determinate matters of fact with
no intrinsic connection to consciousness and its logical operations in
judgment or language, then it seems gratuitous to posit logically complex states of affairs. Indeed, so it has seemed since Wittgenstein's *Tractatus*. Wittgenstein recognized elementary states of affairs (*Sachverhalten*)
of form [aRb] (and arguably situations [*Sachlagen*] of form [aRb and not
cQd]). But he did not recognize quantified states of affairs of form [All
Fs are Gs] or [Some Fs are Gs]. Wittgenstein restricted quantification to

propositions (*Sätze*) or thoughts (*Gedanken*), as in propositions of form "All As are B" and "Some As are B." Now, Whitehead gives propositions a modal status between that of purely "objective" actual entities (faits accompli) and that of purely "subjective" prehending entities: as hybrid entities opening toward prehension, or *proposed for feeling*. But potentialities may be toward "some" or "all" of a set of actual entities, not only particular concrete actual entities, being determined in a certain way. Then it is plausible, surprisingly, that there are in the world "proposed" states of affairs such as the proposition of form "Some Fs are caused by a." This hybrid entity integrates the actual entity a with the potentiality of "some Fs" to be related by causation to a. This "proposition" is just as much a part of the world as is the proposed state of affairs of form "a causes b," which integrates two actual entities with the eternal object causation.

Here is a more robust variety of "logical atomism" than Russell's or Wittgenstein's. For Whitehead's ontology of "propositions" allows *all* the familiar forms of propositions, not only the singular ones, while placing them all in the world alongside concrete states of affairs and not merely in language or thought or its contents. And thereby Alfred North Whitehead puts logic in its place: in the cosmos!

Consciousness

For Whitehead, consciousness emerges, in the order of the cosmos, only with higher phases of prehension: an actual entity is "conscious" just in case it prehends a *proposition* in a certain way (to be specified later). Fundamentally, then, consciousness is a propositional attitude, to use Russell's idiom. A more systematic phenomenology, I believe, shows that not all acts of consciousness are propositional in form: judging or seeing "that this is a lizard" is propositional, but seeing "this lizard" is not.[7] For the moment, though, for the sake of argument, let us grant Whitehead propositionality. What is striking in Whitehead's account of consciousness is the role of his ontology. For consciousness inherits its structure – I would say its phenomenological structure – from the *ontological form* of prehension involved. There lies Whitehead's novel idea.

Consciousness is not Whitehead's central concern, and he would seem to leave the specifics to others – may we say, to phenomenologists in the lineage of Brentano, Husserl, et alii, whom Whitehead seems not to have read, and in other respects to neuroscientists in today's "consciousness studies." Yet Whitehead explicitly addresses consciousness

after he analyzes propositional feeling, and he remarks on consciousness all along the way in *Process and Reality*. In his own idiom Whitehead
writes:

> [C]onsciousness presupposes experience, and not experience consciousness. It
> is a special element in the subjective forms of some feelings. Thus an actual entity
> may, or may not, be conscious of some part of its experience. Its experience is its
> complete formal constitution, including its consciousness, if any. Thus, in Locke's
> phraseology, its "ideas of particular things" are those other things exercising their
> function as felt components of its constitution. Locke would only term them
> "ideas" when these objectifications belong to that region of experience lit up by
> consciousness. (1978: 53)

By "experience" and "feeling," we know, Whitehead means prehension,
and the "subjective form" of a feeling is the way an actual entity receives
the "datum" it prehends. The form of prehension in an actual entity is
"its complete formal constitution," what defines it ontologically. In some
cases an actual entity is "conscious of some part of its . . . constitution."
An actual entity has what we call "ideas of particular things" insofar as it
prehends those things and the pattern of prehensions falls in a region of
prehension "lit up by consciousness."

In short, consciousness is a particular form of prehension. It consists
in a pattern of prehensions found in the formal constitution of some but
not all actual entities. To be conscious of something, to have an idea of
something, is to have this pattern of prehension. And this pattern is the
ontological form of the act of consciousness: the "formal constitution" of
an actual entity that is conscious.

It is important for Whitehead that consciousness is just another form
of prehension, and not something that exists in a world apart from the
physical in nature. As he writes:

> Locke inherited [from Descartes] the dualistic separation of mind from body. If
> he had started with the one fundamental notion of an actual entity, the complex
> of ideas disclosed in consciousness would have at once turned into the complex
> constitution of the actual entity disclosed in its own consciousness, so far as it is
> conscious – fitfully, partially, or not at all. (p. 53)

Here is a dramatic statement of monism in Whitehead's scheme. There is
but one fundamental ontological structure in every actual entity (not two,
one for the mental and one for the physical): the structure of prehension. And consciousness is a specific variation on this one fundamental
structure.

What variety of prehension, then, is consciousness? Whitehead's answer begins after the analysis of propositional feeling:

The nature of consciousness has not yet been adequately analysed. The initial basic feelings, physical and conceptual, have been mentioned, and so also has the final synthesis into the affirmation-negation contrast. But between the beginning and the end of the integration into consciousness, there lies the origination of a "propositional feeling." A propositional feeling is a feeling whose objective datum is a proposition. Such a feeling does not in itself involve consciousness. But all forms of consciousness arise from ways of integration of propositional feelings with other feelings, either physical feelings or conceptual feelings. Consciousness belongs to the subjective forms of such feelings. (p. 256)

The "subjective" form of a feeling is the form of the prehension as it is absorbed, as it were, by the actual entity (pp. 22, 24), and consciousness is a prehension of a proposition. But there is more. Consciousness originates, or begins to take form, in awareness of the yes and no, in affirmation and negation. And its final form involves the forms of perception, imagination, judgement, and the like.

Whitehead finds the crux of consciousness in something utterly fundamental:

Consciousness is the feeling of negation: in the perception of "the stone as grey," such feeling is in barest germ; in the perception of "the stone as not grey," such feeling is in full development. Thus the negative perception is the triumph of consciousness. It finally rises to the peak of free imagination, in which the conceptual novelties search through a universe in which they are not datively exemplified [that is, part of the datum prehended].

Consciousness is the subjective form involved in feeling the contrast between the "theory" [or proposition] which *may* be erroneous and the fact which is "given." Thus consciousness involves the rise into importance of the contrast between the eternal objects designated by the words "any" and "just that." Conscious perception is, therefore, the most primitive form of judgement. (p. 161)

Notice that negative perception is a positive prehension of a negative fact; it is not a negative prehension, or not feeling something.

Why is feeling negation at the foundation of consciousness? Whitehead's motive here lies in his particular conception of eternal objects or forms, and thus the ontology of propositions, and thus the variety of propositional feeling that defines consciousness. An eternal object or form is, for Whitehead, a capacity or potential, and a potential is a tendency that might or might not be realized. Accordingly Whitehead writes:

In awareness actuality, as a process in fact, is integrated with the potentialities which illustrate *either* what it is and might not be, *or* what it is not and might be. In other words, there is no consciousness without reference to definiteness, affirmation, and negation. Consciousness is how we feel the affirmation-negation contrast. Conceptual feeling is the feeling of an unqualified negation; that is to say, it is the feeling of a definite eternal object with the definite extrusion of any particular realization. Consciousness requires that the objective datum should involve (as one side of a contrast) a qualified negative determined to some definite situation. (p. 243)

So consciousness, or "awareness," always involves some feeling of the *contrast* between actuality and potentiality. This is the fundamental *ontological* structure of consciousness: it is a prehension of this contrast. To embellish Whitehead's example (quoted earlier from p. 161): when I have a perception of "the stone as gray," I have an awareness of the stone's actually being determined as gray and therein potentially being not determined as gray (say, if the stone's temperature changed sufficiently to alter its color). Husserl would have said this awareness of potentiality is part of the "horizon" of the awareness of actuality.[8] Presumably, Whitehead thinks this "horizonal" awareness of potential negative facts is characteristic of consciousness but not other types of prehension.

Specifically, consciousness always involves feeling a proposition. And on Whitehead's analysis, the ontology of propositions is that of a contrast between actuality and potentiality, namely, actual entities and forms:

[T]he many components of a complex datum have a unity: this unity is a "contrast" of entities. For example, a proposition is, in a sense, a "contrast." The most important of such "contrasts" is the "affirmation-negation" contrast in which a proposition and a nexus obtain synthesis in one datum, the members of the nexus being the "logical subjects" of the proposition. (p. 24)

When "different" actual entities are formed into a complex, the resulting unity depends on the contrast between the entities combined. A proposition combines actual entities with eternal objects, which contrast with each other precisely because the former are actual and the latter are not. This is an ontologically fundamental contrast: between beings that are realized (affirmation) and beings that are not yet realized (negation). Whitehead claims, then, that consciousness arises most fundamentally in "feeling" this contrast in the constitution of a proposition. The form of

(form + actual entity)◄—— (propositional feeling)——— act of consciousness
    proposition
negation-affirmation

FIGURE 7.4. The form of consciousness.

consciousness is depicted in Figure 7.4, where an act of consciousness is
a type of actual occasion.

Formal and Material Ontology of Consciousness

What else distinguishes consciousness from propositional feeling in
general? I have not found an answer in Whitehead, and I suspect there
is a reason why I have not.

Whitehead's further analysis of consciousness fans out into analyses of
particular types of "higher phases" of prehension: in perception, imagination, belief, judgment (pp. 266ff.). Here, I believe, we should proceed
instead into the rich analyses of these phenomena that populate the literature of phenomenology, starting with Husserl's results on such things,
which hark back to Hume, Leibniz, and others in ways that Whitehead
would approve. In fact, there is a methodological point to be pressed
here.

My point assumes the distinction between formal and material ontology, along lines charted by Husserl. I believe Whitehead uses such a
distinction in practice, without the terminology. Both were mathematicians who turned to a broader canvas, and in this distinction we see the
intellectual habits of the pure mathematician. Roughly, formal ontology
analyzes structure that is invariant, applying to very different domains,
ideally to all domains, whereas material ontology analyzes structure within
a specific domain.

Whitehead approaches the ontology of consciousness through the
analysis of prehension, propositional feeling, and feeling the affirmation-
negation contrast in a proposition. It is in this latter feeling of contrast,
Whitehead holds, that consciousness begins to emerge, that is, as a distinct type of prehension, leading into higher phases of prehension. But
here, I believe, the formal ontology ends and the material ontology begins. For the structure of prehension – even of feeling a proposition and
feeling the contrast it involves – is purely formal, applying invariantly to
all prehension up to that level. But the structure of our own characteristically human forms of consciousness is partly "material," in that the forms
of feeling in our sensory perceptions, our everyday judgments, our desires and volitions are partly conditioned by the special structures of our

peculiarly human world. Here are forms of prehension that are special to beings like us, in our earthly environment, in this "epoch" as Whitehead allows. Perception, judgment, imagination, and emotion have their formal constitution in what Husserl called intentionality and intentional content, but they also have their specific "material" constitution.

For instance, I see objects around me from the perspective of two eyes looking binocularly toward things I may touch with my two symmetrically placed hands, and some of these things are fellow human beings who share my values and ways of walking and talking and thinking and seeing. There is much to be said about how these "material" ways of prehending (or intending) realize the formal ways of prehending that Whitehead has analyzed. Whitehead the cosmologist has his eyes on the formal structure of prehension, with much less to say about the material structures of perception, imagination, and other processes.

Reading Whitehead in this way, the formal structure of consciousness is that of feeling a proposition and the contrast it involves, whereas the material structure of, say, perception or imagination is that of feeling a nexus of actual entities in a region of space-time.

Reflexive Awareness?

The formal structure of consciousness seems to require one further feature. For an act of consciousness is not only a consciousness of X – a Whiteheadian proposition (we assume for present purposes). It is also *eo ipso* a consciousness *of itself*, of its being a consciousness of X. Without this *reflexive* awareness of itself, the act would not be conscious. Along these lines Brentano and Husserl held that every act includes a primary intention directed toward its object and a secondary intention directed toward itself – but in such a way that no infinite regress ensues (with a third directness toward the second, etc.). Alternatively, I have argued, the reflexive character is a modification of the intention of X, rather than a supplementary intention of the intention-of-X.[9] (See Chapter 3.)

Whitehead does not explicitly address this reflexive structure in consciousness, but he comes close to ruling it out on formal ontological grounds. An actual entity is in process of becoming, and the final phase of its concrescence Whitehead calls "satisfaction." But the structure of concrescence, Whitehead argues, imposes a certain limitation on consciousness: "No actual entity can be conscious of its own satisfaction; for such knowledge would be a component in the process, and would thereby alter the satisfaction" (1978: 85). Whitehead's argument here is elliptic. Becoming cannot be continuous, resolvable into smaller and smaller becomings, Whitehead has argued with reference to Zeno (p. 35). Similarly, the satisfaction of an entity – the final phase of its becoming – cannot be resolved into smaller and smaller component satisfactions. Here the argument may be that satisfaction cannot be resolved into successively nested satisfactions. That is, if an actual entity were to prehend its own satisfaction, then its satisfaction would require the satisfaction (through prehension) of its own satisfaction. With Zeno in mind, this begins to look like a vicious regress.

Whitehead seems to be propounding a formal ontological principle of complete generality: an actual entity cannot prehend its own satisfaction. If we look to the special case of consciousness, however, we find ourselves in more familiar territory: an act of consciousness cannot be conscious of its own becoming. And then Whitehead's principle rules out the reflexive structure that our consciousness seems intuitively to have.

But perhaps we are looking for consciousness in the wrong places. Given Whitehead's "atomism," the things we know in everyday life are not simple "actual entities," but highly complex groupings or "nexuses" of simple actual entities. Perhaps our everyday conscious experiences are such nexuses. Thus, when I see a dog, my conscious perceptual judgment that-this-is-a-dog is a nexus that combines two prehensions: my *perceptual* prehension of the proposition that-this-is-a-dog and my *reflexive* prehension of that perception. These are distinct prehensions, neither of which includes a prehension of itself, of its own becoming. And my conscious perception consists of this nexus. Moreover, the dog and I are ourselves a complex nexus of actual entities. Indeed, Whitehead had his eye on basic physics, and the philosophy of physics must explain the correspondence between basic physical states and everyday affairs.

Husserl's phenomenology of time consciousness, in fact, elaborated such a complex form of awareness. On Husserl's analysis, my current phase of perceptual consciousness includes not only a perception of X but also a spread of "retentions" of successively prior phases of the ongoing perceptual consciousness, and indeed a spread of "protentions" of anticipated upcoming phases of perceptual consciousness. My current ongoing experience thus consists of this complex. And my reflexive awareness of the ongoing experience, we might say, consists in the complex of retentions and protentions. But, as Whitehead requires, no "point" awareness is an awareness of itself.[10]

There remains one further wrinkle by which Whitehead's ontology might capture what we experience as reflexive awareness of experience.

proposition X ◄——— act B ◄——— reflexion

FIGURE 7.5. Reflexive awareness.

Whitehead allows for the "mediation" of feeling through an actual entity (p. 226): if A feels B (in A's actual world), and B feels C (in B's actual world), then A feels C not directly but mediately. Thus, if B is conscious of a proposition X and A prehends B, then A has a mediate prehension of X. Perhaps then A's prehension of B together with its indirect prehension of X constitutes a "reflexive" awareness of B's consciousness of X. See Figure 7.5. Then the act B does not include a component prehending itself, but is rather accompanied (immediately afterward) by a further "reflexion" that prehends it. [Consider the levels and biases of consciousness distinguished in Chapter 3.]

The remaining problem is how well the phenomenology fits with the ontology.

Critical Reflections

We have sketched on Whiteheadian lines a fundamental ontology and its application to consciousness. I should like to close with a brief appraisal of the chief virtues and problems I see in this Whiteheadian system.

The Structure of Consciousness

As Husserl observed, in the wake of Brentano, conscious experience is paradigmatically intentional: as cognitive science puts it today, mind is representational. Some states of consciousness (like feeling dizzy) are not intentional, but most are. Whitehead captures the structure of intentionality in "prehension." So far, so good. He then analyzes consciousness as prehension of a proposition. But while it is often assumed that intentional states are essentially propositional in form (like believing that such and such is the case), some intentional states (like seeing this dog) are not propositional, and so cannot consist in prehending a proposition. Whitehead touches on specific forms of consciousness including perception, judgment, and imagination, but the details do not really further his basic analysis, and we must look elsewhere for the cognitive or phenomenological structures of perception, thought, desire, and the like. What we should learn from Whitehead about cognition is that we need a truly fundamental ontology – in some sense a formal ontology – in order to articulate the structure of cognition and specifically consciousness.

The Ontology of Propositions
On Whitehead's analysis, consciousness begins to emerge in "feeling" a proposition: a hybrid of actual entities and form (say, that a R b), where the form (the relation R) is a "potential" for definiteness in the actual entities (a and b). Whitehead may have had in mind a sort of superposition of phenomena as envisioned in quantum mechanics. Alternatively, what is potential is the *possible* state of affairs combining these things (that a R b), which consciousness may prehend indirectly through a propositional meaning entity (the thought that "a R b"), which itself comes into being with the act of consciousness. Whitehead's intriguing conception of propositions, somewhere in between actuality and potentiality and thereby opening into consciousness, should be parsed, I suspect, into different ontological structures: actual entities, forms, potentiality, and meanings.[11] In any event, all these things will be subject to the "process" of becoming in Whitehead's overall ontology.

The Ontology of Consciousness
The "hard" problem of consciousness, it is said in philosophy of cognitive science today, is how to account for subjectivity: the felt quality of experience, what it is like to have a certain form of experience. What is hard is precisely fitting consciousness, in all its phenomenological glory, into a world of particles in the void. Whitehead took modern physics to heart, embracing relativity and quantum phenomena, and he developed a metaphysical vision in which everything consists in interactions of prehension whereby actual entities come into being bearing forms. Consciousness and causality are then two types of prehension, and mind and body find their place as different types of nexuses of actual entities, forms, and hybrids thereof. The details of this way of distinguishing consciousness and causality require, shall we say, further work. In particular, are propositions as hybrids a good idea? In any event, the architecture of Whitehead's ontology is admirable: we must look to *fundamental ontology* to solve the mind-body problem, so that the distinctive structures of mind and body emerge respectively from more fundamental ontological structures something like ontological becoming.

The Principle of Becoming
In broadest strokes, Whitehead's ontology is a radical revision of the most basic assumption of classical Western metaphysics: that the world is fundamentally structured into forms and substantial particulars related

by exemplification. Plato and Aristotle, and nearly all since, began with this assumption, while differing on the status of the forms. The traditional languages of logic – including quantified predicate logic in *Principia Mathematica* and more recent modal logics – reflect this metaphysics of substances and attributes (including relations). According to Whitehead's ontology, however, all this structure emerges from a more fundamental structure of becoming, or concrescence, through prehension. The principle of ontological becoming, or coming to be an entity, I submit, carries Anaximander's assumption that the order of "determinate" beings – what Aristotle would later call substances (*ousia:* beings) and their species and attributes – emerges from a different order of the "nondeterminate" (*apeiron*). Whitehead begins with such an assumption in the category of the "ultimate," becoming. However, he does not treat substances and attributes in the same way, as "determinates" formed by concrescence. For although he treats Aristotelian substances as products of concrescence, he leaves Platonic forms as "eternal" objects (albeit potentials). This disparity between everyday objects and their forms calls for further evaluation.

The overarching lesson from Whitehead, then, is to see how this principle of becoming might ground all ontological structure. The structure of consciousness is then to be a special case.

Notes

1. Here I refer to the PACIS system developed in different phases, by Ontek Corporation in Laguna Hills, California. PACIS is a high-level database system, designed to represent all kinds of entities in a unified way. I think of PACIS, philosophically, as a computational phenomenological ontology, a system of fundamental categories of ontology (what there is) and phenomenology (how we experience or represent what there is) expressed in its own computer language (how we codify our representations of what there is). The ontology has been developed by a team consisting of the present author, Charles W. Dement (Ontek Corporation), and Peter Simons (University of Leeds), with help from Stephen DeWitt and John Stanley (Ontek Corporation).

2. The dynamicist approach to mind is presented in Port and Gelder 1995. As the editors write, "Dynamicists from diverse areas of cognitive science share more than a mathematical language; they have a common worldview" (p. vii). Their world view, I suggest, is Whiteheadian. That is to say, the greatest classical metaphysical statement of the dynamicist world view is found in Whitehead, with precursors tracing to Hume and all the way back to Heraclitus.

3. The distinction between formal and material ontologies is explicated, in my own terms, in D. W. Smith 1995. I do not here restrict myself to the details of Husserl's particular conception of the distinction, as used in Husserl's categorial ontology. Rather, I want to unfold a version of the distinction framed within Whitehead's very different categorial ontology. Whitehead did not draw the distinction himself, nor had he read Husserl (to my knowledge). However, as a mathematician, like Husserl, he would have implicitly understood the distinction. For it is common to think of pure mathematics as developing formal theories that can be applied to very different domains. For instance, the theory of differential equations may be applied to the motion of waves of water in fluid mechanics or equally well to the flow of waves of radiation in electromagnetism theory. Formal ontology, however, is not concerned with formal structures of mathematical language that *describe* different domains. Rather, formal ontology is concerned with formal structures of reality that are *realized* in different domains.

4. Mayr and Ashlock 1991: 256. The PACIS ontology has been influenced by the search for categorial complexity in biological systematics, although the goal of PACIS is that of a "formal" ontology rather than that of a "material" science like biology.

5. The extant ontology of PACIS develops this ontological architecture, in a particular system of categories, with an eye to both Anaximander and Whitehead. The PACIS scheme, and its Anaximandrian form, are expounded in D. W. Smith 1997. There are important differences between the PACIS system and the Whitehead system. I note one problematic feature of Whitehead's treatment of forms: insofar as forms are prehended, they are entities; but insofar as forms are "pure potentials" for determinateness, they seem to be modes or ways of becoming. The PACIS system clearly distinguishes these phenomena.

6. See Dretske 1981.

7. The case for nonpropositional intentional acts is detailed in Smith and McIntyre 1982 and D. W. Smith 1989. The former book integrates a Husserlian model of intentionality with Jaakko Hintikka's model of propositional attitudes: compare Hintikka 1969. For Hintikka, intentionality is analyzed as propositionality, so that an attitude of belief is in effect directed toward a range of possible states of affairs or "worlds" (compatible with the content of belief). Such doxastically possible states of affairs resemble the "potentialities" that Whitehead speaks of. Even if seeing a lizard is not a propositional act, though, it may be analyzed as embracing a "horizon" of possible states of affairs in which a lizard is potentially before the perceiver.

8. See Smith and McIntyre 1982 for a detailed reconstruction of Husserl's notion of horizon.

9. This reflexive structure of consciousness is studied in D. W. Smith 1989: chap. 2.

10. See Miller 1984 for a precise reconstruction of Husserl's analysis here. On the proposal I am noting, with an eye to Whitehead, what we experience as a reflexive awareness *within* the act of perception is analyzed into smaller "awareness" in the form of retentions and protentions. The problem for this

proposal is that we really are not aware of these distinct phases of retention and protention, and so the reflexive character that we experience in conscious perception is analyzed into "awareness" that we do not actually experience. These issues are discussed in Chapter 3.

11. The alternative model of intentionality is developed in Smith and McIntyre 1982, integrating a Husserlian model of intention via sense and a Hintikkian model of propositional attitudes as unfolding an array of possible states of affairs or "worlds." Compare Hintikka 1969. My concern here is how to understand what Whitehead means by "potentials," if not something definable in more recent modal ontology. In the PACIS ontology, sketched in D. W. Smith 1997, intentional contents or meanings are treated as entities that are dependent on acts of consciousness and do the work of "intending" entities within consciousness.

References

Bergson, H. 1991. *Matter and Memory*. Translated by Nancy Margaret Paul and W. Scott Palmer from the French 5th edition of 1908. New York: Zone Books. French original, 1896.

Dretske, F. 1981. *Knowledge and the Flow of Information*. Cambridge, Mass.: MIT Press.

1995. *Naturalizing the Mind*. Cambridge, Mass.: MIT Press.

Hintikka, J. 1969. *Models for Modalities*. Dordrecht: D. Reidel.

Mayr, E., and P. D. Ashlock. 1991. *Principles of Systematic Zoology*. 2nd ed. New York: McGraw-Hill.

McTaggart, J. M. E. 1927. *The Nature of Existence*. Vol. 2. Edited by C. D. Broad. Cambridge: Cambridge University Press.

Miller, I. 1984. *Husserl, Perception and Temporal Awareness*. Cambridge, Mass.: MIT Press.

Port, R. F., and T. van Gelder, eds. 1995. *Mind and Motion: Explorations in the Dynamics of Cognition*. Cambridge, Mass.: MIT Press.

Russell, B. 1971. "The Philosophy of Logical Atomism." 1918. In Russell, *Logic and Knowledge: Essays, 1901–1950*, edited by Robert C. Marsh. New York: Capricorn Books, G. P. Putnam's Sons.

Smith, B., and D. W. Smith. 1995. *The Cambridge Companion to Husserl*. Cambridge: Cambridge University Press.

Smith, D. W. 1989. *The Circle of Acquaintance*. Dordrecht: Kluwer Academic Publishers.

1995. "Mind and Body." In B. Smith and D. W. Smith, eds., *The Cambridge Companion to Husserl*, pp. 323–93. Cambridge: Cambridge University Press.

1997. "Being and Basis." Unpublished manuscript, extending my presentation under the same title at the Second European Congress for Analytic Philosophy, Leeds, England, 5–7 August 1996.

Smith, D. W., and R. McIntyre. 1982. *Husserl and Intentionality*. Dordrecht: D. Reidel.

Whitehead, A. N. 1978. *Process and Reality*. Edited by David Ray Griffin and Donald W. Sherburne. New York: Free Press, Macmillan. First edition, 1929.

Wittgenstein, L. 1994. *Tractatus Logico-Philosophicus*. Translated by D. F. Pears and B. F. McGuinness. London: Routledge. German original, 1921.

8

Basic Categories

Abstract. This essay pursues basic ontological categories. The strategy is to develop a sense of *systematic* ontology by crafting a series of increasingly sophisticated *category schemes.* We move from an austere scheme of physical particles and sets to the traditional categories of substance and attribute, to a modern view of modality and intentionality, to the distinction between formal and material categories, and various candidates thereof, to a notion of categorial "depth." Along the way we keep an eye out for where mind or consciousness falls in each category scheme. This study indicates that a piecemeal ontology will not deal adequately with basic ontological structures, including the place of consciousness in the world. The methodological lesson to be learned is that the systematic organization of categories is crucial to the practice of ontology.

Segue. In previous essays we explored structures of consciousness and different types of ontology that might help to account for features of

Parts of this chapter follow a series of four lectures I gave at the 1997 Bolzano International Schools in Cognitive Analysis at the conference "Categories: Ontological Perspectives in Knowledge Representation," Bolzano, Italy, 15–19 September 1997. The chapter also reflects a series of two lectures I gave as "Where Mind Meets World" in January 2001 at the University of Miami. This chapter benefited from the discussion on those occasions. In the background lie eighteen years of invaluable discussion with colleagues in the Ontek research program in computational phenomenological ontology: my thanks to Charles W. Dement, Peter Woodruff, Peter Simons, and Steve DeWitt, for wide-ranging discussions of grand metaphysical issues.

See D. W. Smith 1997 on the ontology of "being and basis," featuring basic "modes of being." See Simons 1994–95 with a partial list of the categories, which I call "modes of being." Both of these presentations sketch aspects of the system of ontology developed within the research program at Ontek Corporation in the early 1990s. Much of the groundwork for that system was developed in my weekly discussions with Charles W. Dement of Ontek. The conception of ontological systematics evolved within the research at Ontek in the 1990s.

consciousness including its intentionality and its dependence on both neural activity and cultural background. In this essay we turn exclusively to the problem of developing a detailed category scheme that allows us to deal with a variety of structures of the world, including the fundamental ontological differences among causal relations in nature, intentional relations in consciousness, social relations in culture – and ultimately with the organization of very basic categories or modes-of-being, including dependence, intentionality, unity, and process.

In this chapter we explore a series of category schemes drawn, with creative interpretation, from prominent systems of ontology discernible in Quine, Aristotle, Descartes, Husserl, and Whitehead. The point is to assess the categories and architectures in these schemes by studying the systems in proximity to one another. Along the way, we note where mind or consciousness would fit into each scheme, keeping an eye on where phenomenology intersects with categorial ontology.

This series of category schemes looks toward the ideal of a comprehensive, up-to-date system of categories sensitive to the most basic structures we find in the world around us, including our own conscious experience. From this study we hope to learn something about ontological "systematics," about the organization of categories in a basic ontology.

The notion of ontological categories once seemed to me the least illuminating part of philosophy: a mere list of broad kinds of things. However, by comparing different category schemes, with an eye to concrete things in everyday life, we may come to realize that the *organization* we posit among basic categories sharply defines the form of the world as we know it. That strangely abstract level of reflection, I think, we need in philosophy today, not least in order to reassess naturalism and our theory of mind. The explorations here are meant to proceed in this spirit of developing a greater sense of system in ontology.

Systematic Ontology and Category Schemes

In biology *systematics* is defined as "the science of the diversity of organisms."[1] Systematics classifies organisms into species and larger (and smaller) taxa, ranks taxa in categories, traces the evolutionary descent of species, and decides what determines the identity of a species, a taxon, a category, and so on. According to current evolutionary biology, each species has a treelike path of descent from other species; this path of phylogenetic descent is called its "clade." There have been vigorous debates about whether a species is defined properly in terms of phenotype (observable characteristics), genotype (genetic constitution), molecular constitution, or clade and about what determines various biological groups and their organization. In biological systematics the term "category" is used not simply for important groups or taxa, but for "rank in a hierarchy of levels to which taxa are assigned, such as subspecies, species, and genus."[2] Thus, there is far more to biological systematics than a listing of species: the definition and organization of groups is crucial, even in determining what makes a group of organisms a species.

Now, philosophy needs to develop its own *ontological systematics*, which we may define as the theory of the diversity of basic kinds of entities in the world – not just living things but anything at all. Such a "science of diversity" is what this essay is exploring.

A *basic ontology*, let us say, is a theory about the most basic structures of the world, of what there is, beginning with the most basic divisions among things in the world. Within traditional philosophy, basic ontology begins with basic *categories* of entities, and in the idea of a scheme of categories we find philosophy's first approach to an *ontological systematics*. The explicit conception of ontological categories began with Aristotle, who posited the categories of "substance" and various types of "attribute." Many philosophers implicitly assume categorial distinctions, as between universal and particular, or between mind and body, or between fact and value. Ontology becomes systematic, however, when it organizes such distinctions, seeking a unified system of divisions that define very basic kinds or categories of entities. A *system* or *scheme* of categories organizes such fundamental divisions in the world, somewhat as biological systematics organizes fundamental divisions in the biosphere.

Roughly, ontological *categories* are abstract kinds defining entities with fundamental roles in the structure of the world. Logically, they are designated by general terms or common nouns such as "individual," "attribute," "relation," "event," and "causation." It is the task of basic ontology to specify categories and their roles in defining the structure of the world. We use terms like "entity" and "thing" in the widest possible sense to cover anything in the world, including whatever falls under basic categories – things as diverse as relations and events, and not merely palpable objects like stones or sticks or birds or people.

A well-formulated basic ontology will include a *scheme of categories* of things in the world plus a *system of principles* about entities in the categories (and how they relate to each other, if they do). If a theory is articulated as a formal axiom system, it is launched by a list of terms and a list of axioms, from which logic permits the deduction of theorems.

In principle, a basic ontology may be expressed in the style of a formal theory, with a list of category words and a list of axiomatic principles. In practice, however, ontologies (their categories and their principles) are usually expressed informally, not least because we must capture intuitions whose axiomatization is itself in question, but also because we do not understand categories initially – naturally and intuitively – in a formal axiom system. An axiom system is something we aim for when we seek a certain type of precision. In any case, whatever the style of expression, a category scheme is concerned to map out the most basic divisions among things in the world, divisions taken for granted in other ranges of theory. Principles about things falling in these divisions then ensue.

Because categories reflect the most basic divisions among entities, they are the most basic part of the structure of the world, and so specifying a system of categories is the most basic part of an ontology. However, what counts as a category will itself vary with the system of categories. For how we divide the world into basic categories of things-in-general will depend on how we think the world is structured, even at the basic level of categories. Is the world entirely composed of particles in physical space-time? Is it composed rather of ideas in many minds or perhaps in one great mind? Or is it composed, as some seem to imply recently, of evolving modes of discourse or cultural practice, from the linguistic to the political? Or is it composed entirely of temporal process itself, rather than the discrete things we encounter in everyday business, either physical objects or ideas or political institutions like traffic laws? What types of distinctions among such things define *categories*?

When empirically minded philosophers and scientists attack "metaphysics," it is usually on the assumption that metaphysics is an a priori discipline putting forth speculations that precede and outrun all empirical observations, even in principle, or at least all current scientific results. Is basic categorial ontology a priori and speculative in that way? I think not. We begin with a large base of empirical knowledge, ranging from our everyday, commonsense knowledge of things around us to our best current science. The task then is to ferret out what seem to be the most basic categories assumed, often implicitly, in that knowledge base. And then the task is to organize these categories in a systematic way, thus characterizing the basic structures of the world to the best of our current knowledge. Categorial ontology, carried out in this way, is not blind speculation from the armchair; it is a form of *theory* in the best sense, abstract theory about what there is in the world, but it is an abstraction from empirically gained knowledge, spread out from common sense into science.

Basic ontology is broadly empirical in this way, or a posteriori, because we all have acquired a mass of knowledge from our culture and our own experience long before we can begin to write out an ontology. Still, basic ontology, beginning with categories, is not simply generated by making observations or running experiments. Rather, with a range of experience and experiments under our belts, we begin to fit an organization – a structure of patterns – onto our knowledge of what there is in the world around us. And this task is the proper business of ontology, as distinct from either empirical science or journalism or humanistic interpretation of cultural phenomena. Somewhat as physicists cast about for appropriate mathematics to organize their results into a proper theory, so ontologists cast about – at a high level of abstraction – for different categories and organizations of categories. In this way ontology is broadly a priori. But then ontology is to one extent a priori and to another extent a posteriori. As W. V. Quine has characterized knowledge formation, there is no sharp distinction between those elements of knowledge that we might call a priori and a posteriori. The more a priori principles concern abstract forms, including mathematical and logical forms, but also basic ontological structures and categories. And the more a posteriori principles concern more concrete forms of entities we encounter, to begin with in sense perception.[3] The pedagogical problem for the philosopher in modern culture is that ontology is often a more abstract kind of theory than we are used to, even in mathematical physics. (Mathematical language is not the only measure of abstract thinking; witness the poetic talk of "being" in the contemplative disciplines.)

Methodology aside, the best way to understand what is basic in ontology, and what constitutes an ontological category, is to study a variety of category schemes.

Particles and Sets (Quine)

We begin with a lean and modern category scheme, abstracted from the philosophy of Willard Van Orman Quine (1908–2000). This scheme will serve as a foil to the more complex schemes we address along our way.

W. V. Quine has been the most influential of recent thinkers carrying forth the philosophical legacy of modern logic (since Frege, Whitehead and Russell, Gödel, Tarski, et al.). Quine has put forth an extremely economical ontology reflecting a basic naturalism informed by the methods of logic in service to mathematics, itself in service to physical theory. To parody:

The universe is composed of physical particles.
Their behavior is described by mathematics.
Mathematics is reducible to set theory plus logic.
Logic consists in transformations on sentences, which are patterns of sound, ink, or information bits in brains or computers, all composed of physical particles.
So all there is in the world are *particles* and *sets*.

Quine's actual views are more nuanced.[4] But for our purposes here, let us address the austere variant of Quinean ontology, which is nearly a commonplace world view for many philosophers today who seek hard-nosed science governed by exact mathematics in accord with precise first-order logic.

On the received view, all the sciences rest on physics, as our overall physical theory builds from particles in fields to atomic and molecular structures, to chemical interactions, to biological activities and evolution, to psychological and social phenomena in biological organisms and species, and on up the scale to cosmological phenomena.[5] In short, today's naturalism holds that the universe in its entirety is composed of physical particles in physical fields: in a gloss, "particles in the void" – as the atomists of ancient Greece guessed. What distinguishes today's physical ontology from the ancient (besides a lot of experimental detail) is mathematics: the laws of gravitation, electromagnetism, quantum wave-particle superposition and collapse, and so on are formulated in terms of mathematical structures (starting with the calculus). And on the standard view, all mathematics reduces to set theory. And all mathematical proof follows logic, which is systematized today in first-order logic. Accordingly, the scientific ontologist structures this view of the world with the Quinean category scheme of particles and sets.

The category scheme in this ontology is remarkably simple:

A Quinean Category Scheme
Categories
1. Particle
2. Set

The great virtue of this scheme is its economy. Ockham's razor has been deftly applied to our contemporary scientific world view. William of Ockham, in the thirteenth century, pared the reigning Aristotelian

ontology down to the lone category of "substance," or concrete particulars. Quine, in the twentieth century, added one category of abstract
particulars – namely, sets – to accommodate modern mathematics.

According to this Quinean ontology, there are no universals (kinds,
qualities, relations) beyond sets and ordered tuples (universals "in extension"). And there are no modalities such as necessity (which apply
to entities such as propositions or states of affairs, whose identity conditions are for Quine too unclear to be acceptable). And there are no
mental events or ideas unless they reduce to the behavior of particles
in neurons or perhaps computers. Nor is there any mental representation or intentionality beyond physical processes in a brain or computer.
If consciousness is to be accommodated in this ontology, it will have to
be defined somehow in terms of patterns (defined in terms of sets) of
physical processes of particles in neurons in brains (or some other extraterrestrial physical processes).

Are there *categories*, then, in the world? If categories are high-level types
(kinds "in intension"), then the Quinean ontology posits no such entities.
Instead, there are two "categorial" supersets: the set of all particles (for
all space-time) and the set of all sets (enter set theory to resolve potential
paradoxes).

The preceding scheme can be seen as a twentieth-century mathe-
matical-scientific heir to Plato's original distinction between the concrete
and the ideal. As we work through more complex category schemes,
we may watch how further categorial distinctions bring in more and
more ontological structure. The challenge for this austere scheme is to
make the case that we can really do without these further ontological
distinctions. I think we cannot, contrary to our neo-Quinean naturalistic
Zeitgeist.

Substance and Attribute (Aristotle)

Categorial distinctions were already at work in Plato, not least in discussions of "the one and the many," as in the *Sophist.* Yet the theory of
ontological categories began officially with Aristotle in the *Categories* (ca.
350 B.C.). The Greek term *kategoriai* originally meant what one can say
against someone in court. Aristotle offered a list of ten "types of predicate," which he sometimes called simply "the predicates," or *kategoriai,*
and which later philosophers came to call the "categories." Thus, Aristotle
redefined the term for his purposes to mean the kinds of things one can
say about something (kinds of predication) and thus the basic kinds of

CATEGORIES	Examples
1. SUBSTANCE	
a. PRIMARY SUBSTANCE [INDIVIDUAL]	this man [Socrates], this horse
b. SECONDARY SUBSTANCE [SPECIES]	man [kind], horse [kind]
2. QUANTITY	
a. DISCRETE	four cubits
b. CONTINUOUS	[much (water)]
3. QUALITY	
a. STATE/CONDITION	grammatical (in the soul)
b. CAPACITY	[rational]
c. AFFECTIVE QUALITY	white
d. SHAPE	[square]
4. RELATIVE [RELATION?]	double, half, larger [than]
5. WHERE [PLACE]	in the Lyceum, in the marketplace
6. WHEN [TIME]	yesterday, last year
7. POSITION [ARRANGEMENT]	lying, sitting
8. HAVING	has shoes on, has armor on
9. DOING [ACTION]	cutting, burning
10. BEING AFFECTED [ACTED UPON, PASSION]	being cut, being burned

FIGURE 8.1. Aristotle's category scheme.

things that can be said to be. The categories he proposed reflect everyday grammar (in ancient Greek) and define answers to familiar forms of question. Each category answers the general question, What is it? and a more specific question, Which is it? What kind? How many? How qualified? Where? When? Aristotle's categories are sometimes misread as grammatical categories, or "predicate" types. However, they are *ontological* categories: Aristotle is focused not on forms of language but on forms of "being."[6]

Aristotle listed ten categories, and several subcategories, with examples of each. His category scheme may be summarized as shown in Figure 8.1 (the organization and examples are his). "Of things not in combination," Aristotle said, there are these ten categories. We address "things in combination" later, including states of affairs. Aristotle said nothing explicit about them in the *Categories*, but his phrase implies a place for them in a complete ontology.

For Aristotle, these ten categories are the most *general types* of things in the world. Every entity falls under one and only one of the categories. Yet primary substances occupy a distinguished niche. For species, quantities, qualities, and the like are each "predicated" of primary substances: that is, a primary substance *is* such and such, for each remaining type of thing. Thus, Socrates *is* a human, *is* one in number, *is* rational, *is* smaller, *is* in the marketplace, *is* at noonday, *is* standing, *is* sandaled, *is* criticizing, *is* criticized. There are accordingly ten types of predication, or *being* such-and-such, for species, quantities, qualities, etc.

The Aristotelian scheme differs from the Quinean by adding to "primary substances" (ultimately, for the Quinean, particles) some ten types of "predicate" or "attribute" (categories 1b and 2–10), all of which the Quinean must explicate in terms of sets or else eliminate. In retrospect, an Aristotelian approach distinguishes different categories expressed by various forms of predicate in everyday language, whereas a Quinean approach follows today's predicate logic in assimilating all predicates to one form ("Rxy") with which is correlated a set or ordered tuple of entities.

The most vexing term in Aristotle's discourse is "substance." The Greek word is *ousia*, which strictly means that which is, a *being* or *entity*. The traditional translation as "substance" is completely misleading in English, but Aristotle's examples make it plain that "primary beings" are simply particular, concrete entities – what we may call concrete individuals. In Aristotle's scheme, concrete individuals form the first and most basic category (or subcategory), that of "primary beings/substances," because they are the kind of entity that most properly has being, or is most fundamental in being. Species are the second most fundamental kind of entity, and so form the (sub)category of "secondary beings/substances." An individual's species is essential to it and constitutes its "real definition"; individuals and species are together the most proper "beings" and so form the category of "being/substance." Aristotle's paradigm here is that of living organisms. In light of evolutionary biology today, it was prescient of Aristotle, the founder of the science of biology, to give such pride of place to species – notwithstanding today's rejection of his essentialist theory of the real definition of species.[7]

If Aristotle's ten basic kinds of entity are proper *categories*, then not only are these ten kinds distinct, but these ten kinds of entity have unique ontological roles and stand in unique ontological relations to one another. Accordingly, Aristotle says that a quality is "in" a primary substance (this whiteness is in this piece of paper), whereas a species is not "in" a primary substance but rather is "said of" it (Man is said of Socrates). In more congenial idiom, we may say that a quality *inheres* in an individual while an individual *belongs* to a species. Then inherence and belonging are distinct types of "predication," which are modes of connection between entities in different categories. To complete the ontology, we would specify the various types of connection between entities in the various categories, especially the types of predication that tie concrete individuals to their species, quantities, qualities, relatives, places, times, and so on. Ontological *dependence* is an important feature of the connection between certain entities of different categories (though Aristotle did not use a term for dependence). Specifically, Aristotle holds, a quality cannot exist apart from a substance in which it inheres, while the substance could exist apart from that quality (as when the substance changes). Thus, a quality is a particular entity, not something shared like a species (or, for that matter, a Platonic form). Husserl would later call such entities "moments," or "dependent parts"; today these particularized properties are called "tropes." By contrast, a species is shared by its members and can exist apart from any one of its members, though an individual cannot exist without its species. It is tempting to think that Aristotle believed that entities of the second through tenth categories could not exist apart from primary substances (and the species to which they belong).

Some of Aristotle's categories look odd or half-baked today (we may lose something in the translation). Why "relative" instead of relation? (Philosophers had trouble with relations until the twentieth century, partly because Aristotelian logic recognized only monadic predicates, whereas modern logic recognizes n-place predicates modeled on mathematical functions with n arguments.) Why add "position" – isn't this a complex relation of elements? What is "having" in the relevant sense, other than a type of relation? Why "passion" in addition to "action" – isn't passion just the inverse of action? Is "action" intentional action or merely causal action? And where is mind in all this? And is intentionality all packed into "action" or is it a special form of relation, or "relative"? Rather than dig further into these details, we shall outline next a more modern version of broadly Aristotelian categories, based on the distinction between substance and attribute, or individuals and their properties, but recognizing further features like dependence and intentionality.

Aristotle ramified his ontology in the *Metaphysics*, with well-known doctrines: the "hylomorphic" principle that a concrete being is matter shaped by form; the account of change wherein a being's qualities (but not species) vary over time; and the theory of cause, divided into four types, namely, formal, material, efficient, and final causes. Also, in *De Anima* (Of Spirit, *anima* in Latin, *psyche* in Greek) Aristotle posited the mind (*psyche*) as part of the form of the body, and there he observed what Husserl (after Brentano) theorized as the intentionality of perception and knowledge. (In perception, for Aristotle, the mind takes in the form but not the matter of the object known.) To complete an Aristotelian ontology, then, we would have to explicate these structures of the world and show exactly how they all play in the category scheme. Specifically, we would have to decide where mind figures in the scheme. Presumably, a mental act is a quality (a condition) in a primary substance such as Socrates, and this quality is shared by the act's object.

Starting from natural language, as Aristotle did implicitly (sometimes explicitly), is not a bad way to begin sifting through our knowledge base in search of what we take to be the most basic kinds of entity in the world. For our language, be it Greek or English or Chinese, is the residue of thousands of years of dealing with the world in very practical terms. But let us remember that we are seeking basic categories of beings in the world around us, not logical or grammatical categories of expressions in our language. That said, we articulate and represent ontological categories, either explicitly or implicitly, only within a system of thought and, realistically, within a language that expresses a theory about entities in the world.

Plus Intentionality and Modality (Aristotle Modernized)

Suppose we wanted to preserve as much of Aristotle's ontology as possible today, while recognizing more modern conceptions of relation, set, part-whole, causation, dependence, modality, and intentionality (all of which Aristotle addressed in some form in works after the *Categories*). I have in mind approaching Aristotle's metaphysics in a spirit like that in a rich study by Montgomery Furth,[8] and then folding in intentionality and modality. The categories we would recognize include forms of complexity beyond "things not in combination."

In this spirit, we might modernize Aristotle's category scheme along the lines shown in Figure 8.2 (deleting some of Aristotle's categories and adding other categories).

In this category scheme, we begin with a variation on the distinction between substance and attribute. "Primary beings," or "substances," are replaced by individuals, including concrete particulars, but perhaps also abstract particulars such as numbers and sets. Individuals are the "substrate" of all "attributes," or "properties" (to choose a slightly more modern term). Properties of individuals are divided (in an Aristotelian spirit) into some four distinct types: kinds or species (sortal properties including natural kinds such as biological species and chemical substances, but also including man-made kinds or artifacts), qualities (nonsortal, monadic properties), and relations (polyadic properties, with plural subjects). Quantities are a different type, aligned with numbers but arguably

CATEGORIES

1.	INDIVIDUAL
2.	SPECIES
3.	QUANTITY
4.	QUALITY
5.	RELATION
6.	LOCATION
7.	INTENTIONALITY
8.	COMBINATION
9.	DEPENDENCE
10.	MODALITY

— where:

6.   LOCATION covers extension in space-time
7.   INTENTIONALITY includes
    PERCEPTION
    THOUGHT
    IMAGINATION
    DESIRE
    VOLITION
8.   COMBINATION includes
    SET
    WHOLE
    STATE OF AFFAIRS
9.   DEPENDENCE includes
    CAUSATION
    (ONTOLOGICAL) FOUNDATION
10.   MODALITY includes
    POSSIBILITY
    NECESSITY
    ACTUALITY

— and where:

Intentionality is a mental act's being directed toward an appropriate entity.
A state of affairs is a combination formed by an individual's belonging to a species or bearing a quality or standing in a relation to one or more individuals (or thinking about something or . . .).
A state of affairs is possible or necessary or actual.

FIGURE 8.2. A Modernized Aristotelian category scheme.

grouped under properties of individuals en masse, or properties of sets (which are abstract individuals). (Various kinds of mathematical entities would need to be appraised, deciding inter alia whether all mathematical entities can be built up from sets; we do not delve into these issues here.)

Entities in the remaining categories – such as locations, intentionalities, combinations – might be thought of as properties with special importance. But here they are instead grouped as something other than properties of individuals. Let us call them "modes," borrowing a late medieval term for properties. Here we are beginning to group these "special" properties in an ontological niche of their own, and we make more of this categorial maneuver in other category schemes considered later. Thus, locations place individuals (and their properties) in space-time. Intentionalities direct mental acts (a kind of individual) toward various entities. Combinations unite entities in heaps, sets, wholes, and the like. Dependencies tie entities together in their existence, either as one event causes another or as a quality depends on a particular in which it inheres. Modalities modify the existence status of entities, with the so-called alethic (truth-related) modalities of actuality, possibility, and necessity. Finally, intentionality is ontologically unique (it is neither a quality nor a relation nor anything else), warranting its own category. Thus, individuals form a substrate for properties of different types, and different types of modes modify entities formed from individuals and various types of properties.

Our purpose in sketching (so tersely) this neo-Aristotelian scheme is to bring out some of the complexity that unfolds in an updated broadly Aristotelian ontology. Thus, we distinguish three *types of categories* in this scheme, reorganizing the scheme as shown in Figure 8.3.

William of Ockham famously rewrote Aristotle's category scheme (ca. 1300) by reducing the original ten categories to one, if you will, the category Individual, "nominalizing" or conceptualizing the rest. The Quinean follows suit but finds need for sets. By contrast, our neo-Aristotelian refines Aristotle by eliminating some categories, adding others, and then *grouping* importantly different categories – a far cry from Ockham's list-of-one. This neo-Aristotelian holds that we are not

A. SUBSTRATE
    1. INDIVIDUAL
B. PROPERTY
    1. SPECIES
    2. QUANTITY
    3. QUALITY
    4. RELATION
C. MODE
    1. LOCATION
    2. INTENTIONALITY
    3. COMBINATION
    4. DEPENDENCE
    5. MODALITY

FIGURE 8.3. An ordered neo-Aristotelian category scheme.

multiplying entities beyond necessity. We are instead articulating ontological distinctions that we *need* in order to account for the complexity we find in the world. Ockham's razor is wielded with different results.

In this latter scheme we begin to recognize categories of a *higher order*. Aristotle assumed a flat list of categories, with subcategories, although Primary Substance has a privileged status. Here we group the categories with a sense of ontological priority (still in the spirit of Aristotle). And we go well beyond a Quinean ontology by positing a variety of nonextensional categories including both properties and "modes" of various types.

We note that mind and consciousness appear in this scheme with intentionality, which does not reduce to physical composition or causal dependence or functional relations or anything else.

Substance-Attribute Applied to Mind-Body (Descartes)

A familiar touchstone in metaphysics is the dualism developed in René Descartes's *Meditations on First Philosophy* (1641). Descartes (1596–1650) assumed a neo-Aristotelian distinction between substance and attribute (or "mode" in Descartes's preferred idiom,[9] a different use of the term than mine just presented). Then Descartes argued for the "real distinction" between mind and body. But the first distinction cuts across the second. This dual dualism leads into an instructive system of ontological categories. For here we find *two types* of category: one including Substance and Attribute, the other including Mental and Corporeal or (as we say today) Physical. Moreover, the first *applies to* the second, so the two types cross to form a "Cartesian product" of the two types (to play on a term from mathematics). Thus, we may find in Descartes's ontology a *two-dimensional* category scheme (entering in the boxes in Figure 8.4 the kinds of entity that fall under the crossed categories). As noted, Aristotle's category scheme was a flat sequence of ten categories, with subcategories. And the structured neo-Aristotelian scheme followed the same architecture, albeit with specially grouped subcategories. But with

CATEGORIES

	A. MENTAL	B. PHYSICAL
1. SUBSTANCE	minds	bodies
2. ATTRIBUTE	thought	extension

FIGURE 8.4. Descartes's category scheme.

Descartes's scheme, as organized here, we find a new kind of architecture in category schemes, where two types of category *cut across* each other. (Here let us assume the Aristotelian and neo-Aristotelian categories are simplified to Substance and Attribute, as our concern is how these categories cut across Mental and Physical.)

Of course, mind is at the root of the architectural innovation. Where Aristotle at best treated mind as a quality in certain bodies (consonant with today's physicalism), Descartes gave the mental its own category as distinct from the physical. However, the point of the reconstruction here is not that the mental and physical form distinct categories. Rather, the point is that there are *two types* of category cutting across each other. If Cartesian substance-dualism is untenable, attribute dualism is nonetheless a step in the direction of diversity among properties, and the moral of the story is categorial complexity.

Most philosophers today reject Cartesian dualism for substances and worry about it for attributes (compare Chalmers 1996). But if I am right, there is something quite different at stake in Descartes's ontology. In retrospect, we may see in the Cartesian category scheme the abstraction of ontological *form* from ontological *content*. Thus we may see the Substance-Attribute distinction as "formal" and the Mental-Physical distinction as "material." In this view, the Substance-Attribute distinction applies to *any* substantive or material ($\neq$ physical) domain of entities, and it happens that Descartes posits two such substantive domains harboring minds and bodies respectively.[10] However, we are getting ahead of our story, applying lessons learned from more recent philosophical movements. The gist of this view is that ontology mirrors logic in discerning *formal* ontological categories, and that is the tack followed explicitly by the early Husserl (nodding to Leibniz and Bolzano).

Fact and Essence and Sense (Husserl)

In Edmund Husserl's *Logical Investigations* (1900–1) and *Ideas* I (1913) we find, along with the new discipline of phenomenology, a sprawling ontology. This ontology can be drawn together, I propose, as a *basic ontology* that accommodates many of the lessons of the previous category schemes. In fact, Husserl (1859–1938) can be read as integrating several prominent lines of thought from Plato and Aristotle onward: the ontology of universals, the logic of sense and reference (modifying Frege et al.), the epistemology of sensory-conceptual knowledge formation (modifying Hume, Leibniz, Kant, et al.), and the theory of intentionality (going beyond Brentano), which is foundational for phenomenology. In Husserl's system, the ontology plays a central role but is interdependent with the logic, phenomenology, and epistemology. I laid out a Husserlian category scheme in "Mind and Body" (1995) and explored parts of it more closely in subsequent essays including "Mathematical Form in the World" (2002b). Here I should like to construct (or *construe*) the architecture of the scheme, its implicit ontological systematics. We have to do the work of systematizing Husserl's diverse categories, because Husserl did not explicitly organize his results.

We begin with Husserl's distinction among "fact," "essence," and "sense" (laid out in opening *Ideas* I). We shall ramify Husserl's category scheme in later sections.

Roughly, the type Fact includes concrete individuals and, presumably, events: these entities take the place of Aristotelian "substances." The type Essence (*Wesen*, from *was-sein*, "what" a thing "is") covers Aristotelian "attributes." And the type Sense (*Sinn*) is drawn from logical theory but used in Husserl's own theory of intentionality: sense includes ideal intentional contents (many of which are expressible in language as "meanings").

We may draw the main outlines of Husserl's system of ontological categories, then, as shown in Figure 8.5 (forewarning that Husserl used the term "category" in a more limited way).

In this ontology, Socrates is or was a concrete or "real" (*reale*) entity in space-time – a "fact" (*Tatsache*), in Husserl's idiom, or, as we might say, a "factual" entity. Socrates belonged to the species humanity. For Husserl (like Plato), this species is an "ideal" (*ideale*) entity not in space-time – an "essence" (*Wesen*, from *was-sein*, what is) or, in the older idiom, a "universal." But for Husserl (somewhat like Aristotle), Socrates is a human if and only if a concrete instance of humanity (an instance of the ideal species) is "in" Socrates: according to Husserl's ontology of parts, this instance is a dependent part, or "moment" (*Moment*), of Socrates. Meanings (sense, *Sinn*) are also "ideal" entities not in space-time, but they are categorially distinct from essences. Where essences are *properties* (or kinds or relations) of concrete individuals (or events), meanings are *contents*

CATEGORIES Examples

1. FACT	concrete entities: in time or space-time
2. ESSENCE	ideal species: universals, eidos, forms
3. SENSE	ideal meanings: intentional contents, propositions

FIGURE 8.5. The basic Husserlian category scheme.

of intentional acts of consciousness. Thus, the role of properties is to *qualify* individuals, whereas the role of meanings is not to qualify or "determine" (*bestimmen*) individuals but rather to *direct* mental acts toward their objects, to "present" or "represent" or "intend" objects "as" experienced. Specifically (to compare with semantics), the proposition (*Satz*) "Socrates is a human" is a propositional sense (*Sinn*) that represents or "intends" (*meinen, vermeinen, intendieren*) the state of affairs (*Sachverhalt*) that Socrates is a human. This proposition is the content (*Inhalt*) of my act (*Akt*) of thinking or judging that Socrates is a human, and it is thus expressed by the sentence "Socrates is a human" (an expression, *Ausdruck*, or sign, *Zeiche*). The proposition in my thinking "intends" the state of affairs, which I thereby think "of" or "about."

Many analytic philosophers today (following Russell) take "propositions" to be made up of properties and sometimes individuals, so that a proposition is in effect a possible or putative ("proposed") state of affairs. This terminology is unfortunate, for it ignores the distinction between the *content* and the *object* of an act of thinking. As often observed, the same object can be represented by different concepts: we can think of Socrates as (under the concept) "the teacher of Plato" or as "the husband of Xanthippe." By the same token, the same state of affairs in the world can be represented or intended by different propositions in various acts of thinking: say, where I think "Socrates argued in the marketplace" and you think "the husband of Xanthippe argued in the markeplace." (Allow that my concept "Socrates" is a type of concept tied to my background conception of Socrates, formed from my studies and not reducible to a descriptive concept "the so and so.") We should thus distinguish ontologically the *proposition* that serves as content and the *state of affairs* that serves as object of a propositional act or attitude. In the preceding Husserlian scheme, propositions and their constituent concepts fall under the category Sense, whereas states of affairs are formed from properties falling under the category Essence and individuals under the category Fact. (Russell used the English term "fact" where Wittgenstein used the term *Tatsache*, for an actual state of affairs [*bestehende Sachverhalt*], with a certain form, whereas Husserl used *Tatsache* for an actual concrete object of any type and reserved *Sachverhalt* for a state of affairs.[11] More on states of affairs later.)

As in the *Logical Investigations*, Husserl took forms of language to express forms of sense carried as intentional content by experiences of judgment and perception underlying speech acts of assertion. Husserl offered a complex philosophy of logic, well aware of the new logic of Frege et alii and the emerging set theory of Cantor et alii. But Husserl dug into the cognate ontology more deeply than other philosophers of logic. In any event, our concern here is with ontology rather than philosophy of logic.

For our purposes, let us assume certain priorities in the preceding category scheme. The most basic things in the world are concrete or "factual" entities in space-time. Essences are essences *of* concrete entities. And senses are senses *of* various entities including concrete entities, their essences, states of affairs (formed from concrete entities and essences), and even senses. So essences presuppose or depend on "factual" entities, and senses presuppose or depend on facts and essences and combinations thereof. (Husserl experimented with several orders of priority. One reading of his "transcendental idealism" says that everything depends on intentional content or sense. These issues we leave aside here, assuming a view of priority closer to Aristotle's.)

Mind has a salient position in the basic Husserlian ontology, drawn into the world with the category Sense. On Husserl's general theory of intentionality, an act of consciousness – a type of "fact" – is intentional, a consciousness "of" something, just in case it has or is correlated with a sense, where its sense prescribes or represents an appropriate object (if such exists). Thus, an act of consciousness "intends" or is directed toward an object *x* if and only if it has a sense that represents *x*. So consciousness enters Husserl's categorial scheme with the category of sense, or intentional contents. Of course, Husserl's theory of intentionality via sense is a hallmark of his work and one of his major innovations. Husserlian phenomenology analyzes consciousness by analyzing the various structures of sense that we find in our acts of consciousness. Phenomenology is, in Husserl, a logic or semantics of experience. In the preceding category scheme, if an act of consciousness (a concrete event) is directed via a sense toward a state of affairs, then this relationship involves entities from all three categories.

Formal and Material Essences (Husserl Ramified)

Husserl drew a crucial and innovative distinction between *formal* and *material* ontology, specifically between formal and material essences. Although the distinction between form and content in language has been familiar in logical and linguistic theory at least since Aristotle, a clear ontological distinction between the formal and material aspects of the *world* took centuries to evolve. In Husserl the distinction is finally explicit. In

Descartes there is a place for the distinction. In Leibniz and Kant there are anticipations, but the distinction is limited to ideas or "phenomena." In Husserl, by contrast, the different levels of experience, content, language, and world are explicitly emphasized, and formal structure itself appears in the world beyond language and thought.

Instructively, I think, the formal-material distinction is "shown but not said" in Ludwig Wittgenstein's *Tractatus Logico-Philosophicus* (1921). Thus, in Wittgenstein's scheme, *form* – he calls it "logical form" – applies to propositions and also to the states of affairs they "picture." Indeed, according to Wittgenstein, the proposition (*Satz*) "Rab" pictures and so represents the state of affairs (*Sachverhalt*) [Rab] precisely because the proposition has *the same form* as the state of affairs it represents. We shall not linger here with the problems in Wittgenstein's picture theory of representation (which he famously abandoned). But those familiar with the *Tractatus* should recognize the all-important role of form in the structure of the world. Now, Husserl mapped out this terrain in greater detail some twenty years earlier.[12]

Whereas Wittgenstein was attentive to details of logical form – from the new predicate logic of Frege et alii – in distinguishing objects, relations, states of affairs, and picturing, Husserl recognized these forms but was far more expansive on how broadly logical form plays across the spectrum of logic, ontology, epistemology, and phenomenology. Here I want to bring out the structure of the rich ontology distributed through Husserl's work.

Husserl sharply distinguished *formal* and *material* essences (see my reconstruction in Smith 1995). That distinction ramifies significantly the basic Husserlian – and broadly Aristotelian – category scheme given earlier, which canonized particulars, universals, and meanings. In fact, the formal-material distinction is the most important metatheoretical notion I want to bring out in this essay.

Adding the distinction (and key entries under it), we extend the basic Husserlian scheme as shown in Figure 8.6. Husserl reserved the term "category" for high-level formal essences, including those listed previously. For our purposes, however, we group all the preceding types of entity as proper *categories* (and subcategories) in a Husserlian ontology. For we are organizing different *types* of categories, featuring Husserl's division of essences into formal and material essences. The ellipsis indicates that the scheme includes some further entries to be considered later.

Material essences, for Husserl, are reached by "generalization," moving from fact to species to genus to the highest level of genus under

CATEGORIES

1. FACT [CONCRETE OR REAL ENTITY]
2. ESSENCE [IDEAL SPECIES, ETC.]
   2.1. FORMAL ESSENCE ["CATEGORY"]
      1. INDIVIDUAL
         a. SUBSTRATE [INDEPENDENT INDIVIDUAL]
         b. MOMENT [DEPENDENT PART]
         c. EVENT [DEPENDENT OCCURENCE]
      2. SPECIES
      3. PROPERTY/QUALITY
      4. RELATION
      5. STATE OF AFFAIRS
      . . .
   2.2. MATERIAL ESSENCE ["REGION"]
      1. NATURE
         a. INANIMATE
         b. ANIMATE
      2. CULTURE ["SPIRIT" (GEIST)]
      3. CONSCIOUSNESS
3. SENSE [IDEAL MEANING (SINN)]
   1. INDIVIDUAL CONCEPT, PERCEPT, . . .
   2. PREDICATIVE CONCEPT, PERCEPT, . . .
   3. PROPOSITION (SATZ)
   4. CONNECTIVE: the meanings "and," "or," "not," "if-then"

FIGURE 8.6. The ramified Husserlian category scheme.

which concrete entities occur. At the highest level of generality are three "regions":

1. Nature, embracing physical objects and events and also psychical states (realized in brains in animal bodies).
2. Culture (or "Spirit," *Geist*, as in *Zeitgeist*, "spirit of the times"), embracing social groups, artifacts, institutions, and values.
3. Consciousness, embracing acts of consciousness and their intentionality (abstracted from their dependencies on the natural and social).

Formal essences are reached not by generalization, moving up a species-genus hierarchy, but by "formalization," moving from fact and material essence to formal essence, or *form*. At the highest level of formalization are the formal essences: Individual, Species, Property, Relation, State of

Affairs, et alia. Now, formal essences apply to material essences *cum* facts, much as logical "form" applies to "content" in the logical composition of propositions. Thus, the formal essences of Individual, Species, and so on pertain to entities under each of the material essences Nature, Culture, Consciousness. Under Nature, for instance, there are concrete individuals – bodies, organisms, events – that have species, qualities, and relations, forming states of affairs in "nature." Under Spirit or Culture there are facts – persons and communities – that have kinds, qualities, and relations, forming cultural states of affairs. And under Consciousness there are facts – acts – having species, properties, and relations, forming states of affairs of acts being intentionally directed toward other things, including natural and cultural states of affairs.

We can now say more about states of affairs. For Husserl, a state of affairs is a "syntactic," "categorial" formation: individuals are joined "syntactically" with properties or relations to form states of affairs. For instance, the state of affairs that Socrates is a human is a formal combination of the individual Socrates with the species humankind; and the state of affairs that Socrates taught Plato is a formal combination of the individuals Socrates and Plato with the relation of teaching. Obviously, this ontology of states of affairs reflects the formal structures posited in Fregean-era predicate logic, and the doctrine is shared by Wittgenstein and other Austrian philosophers of the early twentieth century. Although formal semantics in the twentieth century made do with set-theoretic structures (like models) instead of states of affairs, a state-of-affairs ontology is not outdated: David Armstrong (1996) has recently argued anew for "a world of states of affairs."

Husserl joined this ontology with an epistemological notion of "categorial intuition." According to Husserl, we grasp intuitively the formal *categorial* structures of states of affairs, including predication (Socrates *is* a human) and connection between atomic states of affairs (Socrates is a human *and* he taught Plato). Our concerns, however, are limited to the ontology.

Entities under Fact, Essence, and Sense are linked in distinctive ways. Senses *represent* entities under these categories, whereas material essences *qualify* or "determine" factual entities, and formal essences or "forms" *define* or "inform" and so *apply* to entities under Fact, Essence, and Sense. We begin to see a complex of links among things under the three basic categories. There are patterns of recursion in the scheme, as forms apply to facts and essences, forms have essences, senses have essences including forms, and senses represent facts, essences, and states of affairs. A full development of the ontology would map out these types of linkage.

Remember that Husserl did not assume a dualism, or trialism, of "substance." As I (1995) stressed, the same "fact" – this subject ("I") or this experience – may have aspects or moments (dependent parts) falling under Nature, Culture, and Consciousness. Husserl's notion of moment is a variant on the Aristotelian notion of a particularized quality that exists only "in" a primary substance. Thus, this white in this piece of paper is a dependent part of the paper; it could not exist unless the paper existed. We shall shortly factor dependence out as a distinct formal essence.

Under Sense, we now observe, are such types of sense as Individual Concept, Predicative Concept, Proposition, and Connective (to adopt more recent terminology). These types of sense reflect familiar logical forms, which align at least partly with certain formal essences.

Types of Formal Essence (Husserl Ordered)

Husserl developed a wide notion of formal essence, or *form*, wider than that explained just previously. He merged "pure logic" with formal ontology, under which he placed not only the formal categories of states of affairs and their constituents but also *mathematical* forms of quantity and order and *ontological* forms of part and dependence.[13] With some creative interpretation, we may thus distinguish four types of formal essence.

Adding these distinctions, we may order the Husserlian category scheme in the way shown in Figure 8.7. All these types are discussed in Husserl's texts, but their organization is unspecified. Husserl does not offer an architectonic (say, like Kant's diamond of four groups of three categories in the *Critique of Pure Reason*, 1781/1787.) However, we may develop the *systematics* of Husserl's ontology, I propose, by organizing the categories in the preceding scheme. This architecture is itself a theoretical hypothesis. The aim here is to build on Husserl's results by explicitly positing a certain ordering or ranking of these various categories.

A select group of formal essences in the Husserlian scheme are those we may call *predicative* or *syntactic* essences:

Individual, Species, Property, Relation, State of Affairs.

These essences are *ontological forms* ($\neq$ genera) that mirror logical forms of expressions (names, predicates, and sentences). And Husserl's discussion

CATEGORIES

1. FACT [CONCRETE OR REAL ENTITY]
2. ESSENCE [IDEAL ENTITY]
   2.1. FORMAL ESSENCE OR FORM
     2.1.1. PREDICATIVE OR SYNTACTIC FORM
       1. INDIVIDUAL
           a. SUBSTRATE [INDEPENDENT INDIVIDUAL]
           b. MOMENT [DEPENDENT INDIVIDUAL]
           c. EVENT [DEPENDENT OCCURENCE]
       2. SPECIES
       3. PROPERTY/QUALITY
       4. RELATION
       5. STATE OF AFFAIRS
       6. CONNECTION: *and, or, not, if-then* (in states of affairs)
     2.1.2. MATHEMATICAL FORM
       1. UNITY
       2. PLURALITY
       3. NUMBER
       4. SET
       5. MANIFOLD
     2.1.3. ONTIC FORM
       1. MEREOLOGICAL FORM: PART, WHOLE
       2. DEPENDENCE
     2.1.4. INTENTIONAL FORM
       1. INTENTIONAL CHARACTER
       2. INTENTIONAL RELATION
   2.2. MATERIAL ESSENCE OR REGION
     1. NATURE
         a. INANIMATE
         b. ANIMATE
     2. CULTURE
     3. CONSCIOUSNESS
3. SENSE [MEANING ENTITY]
   1. INDIVIDUAL CONCEPT, PERCEPT, . . .
   2. PREDICATIVE CONCEPT, PERCEPT, . . .
   3. PROPOSITION
   4. CONNECTIVE: "and," "or," "not," "if-then"

FIGURE 8.7. An ordered Husserlian category scheme.

of formal essences (in *Ideas* I) rests on his extensive prior discussion of logical structure (in *Logical Investigations*). Yet these "syntactic" formations are ontological; they are not grammatical categories of expressions (or of senses). Thus, Husserl explicitly talks of the "syntactic" form of *states of affairs*, and he holds that we have a "categorial intuition" of states of affairs and their "syntactic" structure. Whereas Wittgenstein in the *Tractatus* held that we can merely "show" such forms, Husserl held that we can in some way "see" them, or grasp them intuitively, and of course we can "say" or express them in language.

Yet Husserl's conception of logic was wider than, say, that of Frege and Russell and Whitehead.[14] In the Prolegomena of the *Logical Investigations*, Husserl defined "pure" logic as a *mathesis universalis*, suggesting that "mathematical" form is logic's aim. Here is the core of the distinction between formal and material "essence" (drawn in *Ideas* I). Still, mathematical forms such as Number and Set have their own ontological roles, in quantifying and grouping entities in any material region. Husserl took geometry to analyze the material essence of space-time or spatiotemporal things; yet the mathematical form – of a physical shape of a physical object in space-time – ought to be a formal essence. Generally, for Husserl, mathematical forms define a type of formal essence distinct from that of the predicative (and connective) "syntactic" ontological forms that structure states of affairs.[15]

In the second of the *Logical Investigations* Husserl addressed "ideal species" (called "essences" in *Ideas* I), treating them rather like Platonic forms. But if whiteness per se is an ideal species, this whiteness in this piece of paper is a "moment" of the paper, rather like an Aristotelian universal. Husserl borrows here from both Plato and Aristotle. Then in the third of the *Logical Investigations* Husserl analyzed the formal essences Part-Whole and Dependence-Independence, in terms of which a "moment" is defined as a dependent part. He also uses dependence extensively when he considers dependencies among the mental and the physical (as mapped out in Smith 1995).

In the fifth of the *Logical Investigations* and again in *Ideas* I, Husserl analyzed intentionality. Intentionality is part of the material essence of an act of consciousness. Yet, I believe, its form must be a formal essence. On Husserl's analysis, an act's directedness cannot be treated as a proper *relation* to its object: because its object need not exist; because the directedness involves a sense that represents the object, so the act is "related" to its object only "under" that sense; because the directedness itself fans out into various possible circumstances; because the directedness involves a "horizon" of further possibilities about the object "intended." Intentional directedness is thus "relation-like," but it is not strictly a relation, which entails that intentionality has a distinctive form, or formal essence.[16]

Factoring out different aspects of form in this way, we may divide the Husserlian formal essences, or *forms*, into four types indicated above.

1. The predicative or "syntactic" forms reflect logical syntax: Individual; Species, Property, Relation; State of Affairs; Connection.
2. The mathematical forms are studied in branches of mathematics: Number, Set, Manifold, and so on.
3. Part-Whole is treated separately as a formal essence. Then there is the pivotal formal essence: Dependence. Dependence is formal because it applies to entities in different material "regions." But it has a different role than either predicative or mathematical forms. It is presupposed, for one thing, in the formal connection between individuals and relations, forming a state of affairs. We need also to consider other categories that belong in this third group, including Modality.
4. Finally, we take intentionality as formally distinctive, as having a separate type of form, *intentional* form. We assume the form of intentionality does not reduce to the form of relation, or to any other type of formal essence. Hence, intentionality has a unique ontological role.

Distinguishing these various *types* of category, on the heels of the previous category schemes, leads to the larger question of higher-order structure in a category scheme, to which we now turn.

"Deep" Structure in Ontology (Whitehead Interpreted)

In *Process and Reality* (1978/1929) Alfred North Whitehead (1861–1947) developed an elaborate categorial ontology designed to supersede the Aristotelian metaphysics of substance-and-attribute, which has reigned (with variation) for 2,500 years. Whitehead's categories, and their application to mind, are explored in my "Consciousness and Actuality in Whiteheadian Ontology" (2001a), reprinted as "Consciousness and Actuality" in Chapter 7. Here I turn to the architecture of Whitehead's category scheme. What is new in that scheme, as I interpret it, is a "deep" structure in the world at large.

Whitehead's metaphysical vision was shaped by both physics and biology: everything evolves or "becomes" in the flux of matter-energy in space-time, and the universe as a whole is itself an evolving "organism." Thus Whitehead proposed to replace the Aristotelian category of "primary substance" (that which becomes) with the category of "becoming": the process of *becoming* an "actual entity." Substance ontology reflects the subject-predicate structure of grammar – even as developed in the mathematical logic of Whitehead and Russell's *Principia Mathematica* (1910–13). By contrast, Whiteheadian process ontology reflects the mathematical structure of the calculus and fits more complex mathematics like dynamical systems theory developed after Whitehead's day. From predicate calculus to field calculus, we might say, is the formal reformation Whitehead sought for ontology. Importantly, Whitehead does not eliminate everyday objects and their properties, where Aristotle began, but he proposes a different and "deeper" foundation for the category of substantial particulars.

Whitehead starts with Heraclitus's view that all is flux. The primary entities in the world are accordingly something like point-events, dubbed "actual entities" or "actual occasions," in the process of actualizing. Things like Socrates and his house or even his walking or talking or thinking are formed from these more basic entities. For Whitehead, each actual entity "feels" or "prehends" a wide variety of other actual entities and also "eternal objects" or Platonic "forms" (conceived as pure potentialities for being determined in some way). In the process of *becoming*, or "concrescence," *one* actual entity emerges out of the *many* actual entities and forms that it prehends. *Prehension* subsumes three traditional ontological relations: *causation* of one actual entity by another, *consciousness* of something by an actual entity (of the right sort), and *exemplification* of a form by an actual entity. A *nexus* is a fact of "relatedness" wherein an actual entity prehends (in different ways) other actual entities and/or forms. The world is thus a flux composed of very many nexuses in which actual entities prehend many other actual entities and many Platonic forms. More precisely, for Whitehead, "the actual world" of a given actual entity consists of those other actual entities and forms it prehends, from its point of view as it were. So there is not one overarching world, but rather a different world relative to each actual entity. The influence of relativity physics is apparent.

Whitehead is normally read as a "process" philosopher who would reduce everything, in Heraclitean fashion, to temporal flow. But Whitehead distinguishes two types of *flux* called "transition" and "concrescence." (See my reconstruction in the essay cited.) As I would like to put it, transition is the process of *temporal becoming*, while concrescence is the process of *ontological becoming*. Thus, *transition* is the temporal "process" by which one entity emerges from others earlier in time. Time, or space-time, is an "extensive continuum" of these transitions. By contrast, *concrescence*, or ontological becoming, is the ontological "process" by which one actual entity *becomes* – becomes that entity – as it emerges out of the many actual entities and forms it prehends.

In our terms here (drawn from Husserl), transition is a *material* "process" and becoming is a *formal* "process." So becoming is the formal process in which all things become entities, become what they are in relation to all else. Time or space-time itself "becomes" in this primordial sense. As Whitehead remarks: becoming is not temporal; temporality becomes. In this way, for Whitehead, becoming is the ultimate structure of being: the "category of the ultimate."

Adverting to the ancients, I like to see this element of Whitehead as echoing not Heraclitus but Anaximander. Anaximander, essentially the first philosopher in the West, around 600 B.C. proposed that all the world – including the most basic elements (air, earth, fire, water: or strings, quarks, electrons, etc.) – originates in a more primordial level called the *apeiron*, or "nondeterminate": that which is not "determined" or defined with the familiar types of properties. In like manner, Whiteheadian ontology posits a "deep" level of structure – Becoming – that ultimately defines the more familiar levels of structure, including substance and attribute. Everything is thus founded on becoming.

Whitehead organized his ontology around a complex system of categories. He posited four types and levels of categories:

1. The Category of the Ultimate: Becoming.
2. Categories of Existence: of what exists or becomes.
3. Categories of Explanation: of how things are (axiomatic conditions).
4. Categoreal Obligations: preconditions on how things can be (conditions of possibility).

Within this hierarchy, he organized quite a number of categories, as shown in Figure 8.8. Most of the complexities (and oddities?) in Whitehead's category scheme we cannot delve into here. (Where Whitehead's terminology deliberately sounds mentalistic, think instead of hard ontological form.) Yet it is instructive to study the motivation and architecture of this scheme, in contrast with the other schemes we have been studying. To be sure, we have come a long way from the simplicity of the category scheme we drew from Quine, whose 1932 Harvard dissertation (in logic) was directed by Whitehead.

What I should like to abstract from Whitehead is an intuitive sense of how differently "becoming" works in his ontology. Becoming is neither an object nor a property or relation of an object. The distinctions we drew

CATEGORIES

1. THE CATEGORY OF THE ULTIMATE: BECOMING
    (CREATIVITY, CONCRESCENCE, PROCESS)
    – the "principle" by which many become or create one; presupposed by all
    other categories.
2. CATEGORIES OF EXISTENCE
    * 2.1. ACTUAL ENTITY (ACTUAL OCCASION)
      2.2. PREHENSION (FACT OF RELATEDNESS)
      2.3. NEXUS [FACT OF TOGETHERNESS]
      2.4. SUBJECTIVE FORM [FORM OF APPREHENSION]
    * 2.5. ETERNAL OBJECT (FORM, POTENTIAL FOR DEFINITENESS)
      2.6. PROPOSITION [= STATE OF AFFAIRS]
      2.7. MULTIPLICITY (DISJUNCTION-OF-DIVERSE-ENTITIES)
      2.8. CONTRAST (PATTERNED-ENTITY)
    – * Where "actual entities and eternal objects stand out with a certain extreme
    finality."
3. CATEGORIES OF EXPLANATION
    27 axiomatic principles about how actual entities, etc., may become.
4. CATEGOREAL OBLIGATIONS
    9 axiomatic principles about preconditions for existence and explanation.

FIGURE 8.8. Whitehead's category scheme.

among *types* of essence in Husserlian ontology move in this direction, when we separate unity, part, dependence, and intentionality from properties that are predicated of objects, properties such as being a human, being wise, or being a teacher. Whiteheadian becoming is something of a different order altogether. It is a kind of "deep" structure determining what it is to be an entity at all, according to Whitehead's ontology. I am inclined to call ontological forms at this level "modes of being," suggesting ways in which an entity comes to be the entity it is. Accordingly, in Whitehead's system the category of Becoming is more fundamental than the categories of Existence, where we find categories closer to others we have considered. The architecture of the Whiteheadian scheme thus places Becoming at a distinctive and basic level.

Systematics in Category Schemes

A basic ontology, we said, is a theory about the most basic structures of the world, beginning with an ontological *systematics* that posits the most basic divisions or "diversity" among entities in the world, divisions defining the basic categories of the world. Logically, a basic ontology includes a scheme of categories and a system of principles about entities in those categories. But there is more to ontological systematics than a list of categories. As our comparative explorations have been shown, the organization of categories is crucial. Thus, a category scheme *organizes* the basic divisions among things into categories, so that there is a *structure* in the basic division of entities. In the series of category schemes explored here, we see an increase in complexity of structure as we move through the schemes. What is the ontological significance of such complexity?

The Quinean scheme posits only individuals (ultimately particles) and sets: no nonextensional entities, and no bona fide categories or types as opposed to sets. The structure of this scheme is thus very simple. In set theory, to be sure, Russell's theory of types adds complexity: a hierarchy of types of sets. But Russell's theory concerns only types of *sets*; at most it can serve as a partial model of how some ontological types – groups – might be ordered. Also, in the spirit of the Quinean scheme, particle physics may lead into complications of quantum and field structure. But still, in this category scheme, the ontological simplicity of individuals-and-sets remains, even as set theory and particle physics grow more complicated as theories.

The Aristotelian scheme posits a flat list or sequence of ten categories, with branching subcategories for some. The structure of this scheme is still very simple: a list of mutually exclusive types and subtypes. However, Aristotelian categories are bona fide types (not sets). They are the most general types of entities that make up the world, the highest general types, *summa genera*. If Species is granted its own place, as deserved, then the Aristotelian scheme is a list of eleven categories, which are the highest genera of entities in the world, prior to combinations thereof. It is assumed that every noncomposite entity falls under exactly one of the categories: the categories are mutually exclusive and collectively exhaustive. Beyond that, the only principle of order is that the category Primary Substance is most basic, because all the remaining categories cover entities that are "predicated" of primary substances, or concrete individuals. If you will, concrete individuals are first-order entities, and all other entities are second-order entities predicable of individuals.

The neo-Aristotelian scheme varies the categories but retains the basic Aristotelian structure of substance and attribute. Thus, different types of "properties" and "modes" are predicable of concrete individuals, the ontological "substrates" of the world. We also find composite entities, notably states of affairs, that are formed by combinations of individuals with properties (featuring the mode of combination), and which themselves carry modes like possibility. Yet the structure of the category scheme is still a list of categories with subcategories. And every noncomposite entity in the world falls under one and only one category (or subcategory).

The interesting question for Aristotelian category schemes is whether there are different connections of "predication" for the different types of attribute, the different categories after Substance. The simplest view posits one kind of link that ties all types of attributes to individuals. Then we would assume that Socrates *is* (univocally) a human, one, rational, a teacher. A more complex view requires different types of link. Then we would assume there are different *basic ontological ties* for different categories. The ubiquitous "is" then covers a multitude of sins, so that it is one type of thing that Socrates *is* a human in kind, quite another type of thing that Socrates *is* one in quantity, quite another that Socrates *is* rational in quality, and so on. In other words, different types of attribute "do" different things to an individual, so these types of predication are ontologically distinct and should be kept apart. Despite the increase in complexity, I am sympathetic to this second approach. After all, what makes Socrates "be" a human is a matter of belonging to an evolving biological species, whereas what makes Socrates be one is a matter of arithmetic structure or countability, whereas what makes Socrates be rational is a matter of how well his mind (or brain?) works, and so on. What could be more different than species membership, countability, qualifiability? When we get serious about ontology, these are the kinds of difference that matter.

Now, the Cartesian scheme presented here introduces a new principle of organization among ontological categories. The scheme is no longer simply a sequence of categories, with branching subcategories. Rather, there are two *types* of categories, one cutting across the other. To impose the later terminology, the formal categories of Substance and Attribute cut across and *apply* to the material categories of Mental and Physical. Thus, there are two basic types of type: formal and material. What is remarkable, in our view here, is not the familiar material division into mental and physical, for which Descartes is so famous. Instead, whatever one may think of Cartesian dualism of substance and attribute (substance-dualism plus property-dualism), I want to bring out Descartes's implicit distinction between formal and material types or categories. These are second-order types. What we find then in the Cartesian category scheme, cast into our idiom, is a new structure of categories: not a sequence, but a table of rows and columns, a *matrix* of two dimensions.

Because modern mathematics began with Descartes, we might expect that subsequent mathematical visions would reshape the organization of basic categories in ontology, not least when logic itself was tied into mathematics in the late nineteenth century. Indeed, that is what Husserl – mathematician turned philosopher – brought into categorial ontology. Accordingly, in the Husserlian category scheme reconstructed earlier, we find much greater complexity than in prior ontologies. Emerging there is a vision of *higher-order* categories interlinked by distinctive *ontological ties* among entities in these categories – ties ranging from predication to mathematical structuring to part-whole composition to ontological dependence to intentional representation. Let us step back and appraise this complexity.

The ordered Husserlian scheme begins with the distinction among the categories of Fact, Essence, and Sense – requiring that these kinds of entities be interrelated. Then Essence is divided into the categories of Formal Essence and Material Essence – requiring that forms apply to entities in different material domains. Then Formal Essence is divided into some four types of Form – requiring inter alia that various forms define entities differently. So there are several importantly different *types of types* of entities: *categories* of types of entities. These higher-order categories do not emerge from a putatively exhaustive listing of entities posited in a wide-ranging ontology. Rather, they emerge through systematic analysis of quite different types of things: logical structure, mathematical structure, predication of universals in particulars, part-whole composition, existential dependence, and (not least) structures of intentionality. It is in Husserl's appraisal of these diverse things that we begin to discern a scheme of higher-order categories – emerging piecemeal without being so christened. I believe this categorial complexity in Husserl's ontology is something new in the history of philosophy: this kind of ontological systematics, with this degree of organizational complexity. (The story is far from finished.)

For the Aristotelian, we noted, different categories may reflect different types of predication as we move from species to quantities to qualities, and so on. What is remarkable, in the ordered Husserlian scheme, is the complexity of the *links* (can we say "formal relations"?) among entities in different categories.

Consider the following basic principles that begin to articulate the ontological links among entities in the different categories assumed in the Husserlian scheme:

1. Essences are *predicated* of entities:
    (a) of concrete individuals and events ("factual" entities have essences),
    (b) of essences (essences have essences),
    (c) of senses (senses have essences).
2. Material essences "determine" or *qualify* entities that have them.
    (a) Natural essences qualify entities in nature, in space-time.
    (b) Cultural essences qualify entities in society, in cultural systems.
    (c) Consciousness's essences qualify entities in conscious experience.
3. Formal essences "define" or *form* entities that have them.
    (a) Predicative essences form entities in states of affairs.
    (b) Mathematical essences form entities in various mathematical structures, such as numerical order, geometrical order, and sets.
    (c) Ontic essences form entities in dependencies.
    (d) Intentional essences form entities in intentional structures: as subjects, contents, or objects of intentional experiences.
4. Senses "mean" or *represent* entities.
    (a) Acts of consciousness "contain" or "entertain" or *invoke* senses.
    (b) An act of consciousness is directed through its sense toward that object (if any) which is represented by its sense.

The categories of Fact, Essence, and Sense are themselves defined largely by such links. These fundamentally different links among entities distributed among the different categories – the links indicated by italics in the principles in the preceding list – entail a complex structure in the category scheme. Rather than a sequence of categories, then, we find a matrix of several dimensions, as entities of different categorial type link or apply to others in different ways. An exact analysis would be a next step in spelling out the ontology – details far beyond the scope of the present essay.

Whitehead, another mathematician turned philosopher, parsed the structure of the world differently than the preceding. In the Whiteheadian category scheme, however, we find an important architectural principle. The category of Becoming is the most basic, the "deep" structure

of all being, and other types of category are built up from there. Without lingering over the details, we can begin to regroup basic categories
with an eye to our reconstructions of both Husserlian and Whiteheadian
category schemes.

Whiteheadian forms and Husserlian essences group together, allowing subdivisions like those presented here. Husserlian senses are a more
articulate part of consciousness than Whitehead allowed (compare my
"Consciousness and Actuality" in Chapter 7). What I would like to stress
most, however, is the architectural principle. As Becoming is basic in the
Whiteheadian scheme, I submit that certain formal essences are basic in
the ordered Husserlian scheme: some categories are, in George Orwell's
phrase, more equal than others. Of the many categories we have considered, I submit, the most basic include those of Predicative Form (being
thus and so), Ontic Form (Dependence, Composition), Mathematical
Form (Number, Set, . . .), and Intentional Form (Sense, Representation).
These types of form we may see as factoring Becoming: these most basic
forms are "modes of being" that formally define what it is to be an *entity*,
that is, of a type higher than these. (Compare D. W. Smith 1997.) Here in
Whitehead's ontology we see another great leap in categorial complexity,
in ontological systematics. Unlike Husserl's systematics, however, Whitehead's organization and typing of categories are explicit – writing from
the heart of modern logic, cognizant of Russell's type theory, but with an
eye to abstract metaphysics rather than logic and set theory. (A scheme
of basic modes of being is outlined in D. W. Smith 1997. See also Simons
1994–95.)

Whereas Whitehead recognized one form of "becoming" an entity,
defined by a pattern of prehension, I believe we need to recognize fundamentally different types of entity defined by different basic modes of
being. Essences ("eternal forms"), senses ("eternal" contents of thought
or experience), and concrete individuals and events – these importantly
different types of entities are defined differently by basic forms or modes-
of-being. Thus, concrete entities are formed by their place in space-time,
while essences (kinds, properties, relations) are formed by their place
in predication, and senses (intentional contents) are formed by their
place in intentionality (which is not properly a relation). In the next section we consider a category scheme that simplifies the results gathered
here.

What I hope is growing clearer and clearer is that a rich ontology –
an ontology rich enough in its *systematics* to capture the diversity of being
around us – requires a category scheme with a lot of structure ordering

the categorial divisions themselves. Again, we are looking at fundamental ontological divisions at least among predication, composition, dependence, and intentionality.

A Twenty-First-Century Category Scheme

Bearing in mind the ontological complexities considered here, let me propose a simpler category scheme that factors the categorial divisions explored. We distinguish formally distinct types as forms, or formal categories. And we distinguish (differently than Husserl) several materially distinct types drawn initially from everyday life. Within some of these categories, formal or material, we note specific subcategories (presented separately in Figure 8.9, for simplicity). Notice that the categorial niche of intentional content, or meaning (sense), falls under the formal structure of intentionality.

I propose this scheme for future adjudication. There is much to be done to flesh out various parts of the scheme, but I propose the outlines as presented earlier. Most important, for present purposes, are the divisions of formal ontological types. Observing such divisions, we are better prepared to face the most pressing ontological issue of our day: the nature of the connection between mind and body, as we understand

FORMAL CATEGORIES [ONTOLOGICAL FORMS]
    INDIVIDUAL
    SPECIES/KIND, QUALITY, RELATION, STATE OF AFFAIRS
    NUMBER, SET, STRUCTURE, . . .
    PART/WHOLE
    LOCATION [FORM OF SPACE-TIME]
    POSSIBILITY, NECESSITY, (ACTUALITY)
    DEPENDENCE
    INTENTIONALITY
– with subcategories:
    INTENTIONALITY
        SUBJECT, ACT/EXPERIENCE, CONTENT/MEANING,
        OBJECT [OF INTENTION], INTENTION [INTENTIONAL "RELATION"]
MATERIAL CATEGORIES (EVERYDAY KINDS (OF TERRESTRIAL ENTITIES))
    OBJECT, EVENT, PLACE, TIME, PLANT, ANIMAL, HUMAN BEING,
    MENTAL ACTIVITY, ARTIFACT, INSTITUTION
–with subcategories:
    MENTAL ACTIVITY
        PERCEPTION, THOUGHT, EMOTION, VOLITION

FIGURE 8.9. A contemporary category scheme.

it in philosophy and science today. Descartes's ontology notoriously left a
sharp division between mental and physical, a division that leaves philosophy confounded to this day. Husserl's ontology allowed a crucial improve-
ment, making room for the cultural as distinct from both the mental and
physical. Still, all these materially distinct things occur in one world. The
crucial distinctions among the mental, physical, and cultural, we should
begin to see, are formal rather than material. With the preceding formal
divisions we can, I think, better understand the ontology of these material
phenomena ("material" as opposed to formal, not as opposed to mental
or cultural).

In the long run we might cast the most basic categories not as types
of entity but as modes of being that define entities, drawing on the conception of depth ontology prefigured in Whitehead (as interpreted in
Chapter 7). In that spirit we would distinguish two principal formal
categories: Being and Basis. Under Being would fall the various categories
of *entity*, such as Individual, Species, and so on. And under Basis would
fall basic categories or "modes" of *being* that define or "originate" entities
of various categories. Thus, the category Individual would be grounded
in the mode of being distinct (Distinctness), the category Species would
be grounded in the mode of being together (Togetherness), the category
Number would be grounded in the mode of being quantified (Quantification), and so on. However, the category Possibility would itself be a
mode, the mode of being possible. Dependence too would be a mode,
the mode of being dependent. And I am inclined also to cast Intentionality as a mode, the mode of being intentional or directed. (Whitehead's
notion of prehension may be seen in this light.) To detail a category
scheme of this sort would be a very long story, far beyond the scope of
the present exploration. Here we can only gesture toward this approach
to configuring a categorial ontology. (See D. W. Smith 1997 and Simons
1994–95 for brief sketches along these lines.)

                              Three-Facet Ontology

There is an important ontological division among three basic aspects or
facets of an entity: what we may call its *form, appearance*, and *substrate*.
This distinction is elaborated in my "Three Facets of Consciousness"
(in Chapter 1). There I argue that the ontology of consciousness requires this distinction: we need to distinguish – on a categorial level –
the formal structure of intentionality, the way an experience appears in

consciousness (including "what it is like"), and the neural substrate and cultural background of mental states. However, the three-facet distinction is thoroughly general, or rather "categorial." How then does this distinction relate to the preceding category schemes?

Briefly, the distinction runs as follows. The *form* of an entity includes its various types of attributes – using "form" in the wide Aristotelian sense (form plus matter), as opposed here to the narrower Husserlian sense of formal essence. The *appearance* of an entity includes the ways it can be known or, most generally, "intended" in a mental state – as reflected in various types of sense or intentional content that prescribe the entity. Finally, the *substrate* of an entity includes the various types of entities it depends on for its existence – background conditions including physical composition, causal genesis, cultural history, psychological motivation, or whatever, depending on its material type. (The appearance of an act of consciousness begins with "inner awareness" as appraised in my "Return to Consciousness," in Chapter 3. The cultural background of an intentional experience, part of its substrate, is explored in my "Background Ideas," in Chapter 5.)

Now, this three-facet distinction can be grounded in any category scheme that gives pride of place to predication, dependence, and intentionality. For an entity's form includes the attributes or essences (of whatever type) that are *predicable* of it, whereas an entity's appearance includes the senses (of whatever type) that *represent* it in an intentional state; and an entity's substrate includes the conditions or entities on which its existence *depends*. Any entity participates in these ways in predication, dependence, and intentionality. So the three-facet distinction applies categorially to any entity – in a world with predication, dependence, and intentionality.

The Husserlian category scheme, notably the ordered Husserlian scheme fashioned here, grounds the three-facet distinction. Indeed, that scheme was constructed precisely with these, among other, issues in mind. And the contemporary category scheme of the previous section similarly grounds the distinction. Again, this simpler scheme was fashioned precisely to distinguish *formal* categories of:

Individual
Species/Kind, Quality, Relation, State of Affairs
Number, Set, Structure, . . .
Part/Whole

Location [Form of Space-Time]
Possibility, Necessity, (Actuality)
Dependence
Intentionality

The three-facet distinction can be seen, then, in this ontology as the beginning of a systematic categorial ontology, an ontological systematics. But we do not stop at *three* basic facets; we parse as many as our ontology recognizes.

To generalize, or rather to formalize, the contemporary category scheme outlined here entails that any entity has distinctive ontological *facets* of:

attribute
mathematical structure
composition
location (if concrete)
modality
existential dependence
intention (intentional representation)

These categorially distinct ontological *facets* of an entity are instances of precisely the several categories posited in that scheme. Importantly, not every entity admits of facets of all these types. Properties or essences do not have location, nor do mathematical structures, nor do senses. Only minds achieve intentional representation. (Language is infused with mind, and vice versa, and so achieves intention.) Some things depend for their existence on others, whereas others do not. Of course, these differences are precisely what lead us to basic categories, basic ontological forms.

Observing such categorial differences is the lifeblood of ontology, no matter the domain of application.

Complexity

Quine declared a famous preference in ontology for "desert landscapes" rather than jungles, meaning he sought simplicity.

However, to vary the metaphor, the Amazon rainforest has spawned more life forms than any other place on Earth, and we had better acknowledge their diversity. For that matter, the desert biosphere is far from simple; get down on your knees and you see all kinds of life forms that you could not observe from a human's eye level. In any event, I want to close the present chapter with a plea – against the *Zeitgeist* – for complexity in ontology. Much as biology must mark divisions among kinds and order the divisions, so must ontology mark basic divisions and order the resulting basic kinds or categories. The closer we look at things around us, from physical to biological to cultural phenomena to – perhaps the hardest to observe – our own consciousness, the more distinctions we must draw. Moreover, many of the basic distinctions we draw are formal rather than material.

But why does the Quinean posit only individuals, and sets of them? Because modern predicate logic reflects a simple ontology: here lies the driving background idea. (Compare the influence of such conceptions according to my "Background Ideas," in Chapter 5.) As language is structured into terms and predicates only, combined by sentence connectives and quantifiers, so it seemed to Quine the world is structured into individuals and sets only. To add Husserl's distinction, a Quinean recognizes only these two formal categories: all further distinctions in ontology are left to material ontologies, which in turn are established by the special sciences, which take over where formal ontology leaves off.

What we learn from these reflections on increasingly complex category schemes, I believe, is that basic ontology needs to mark out a rather complicated order of basic types of entities, both formal and material types. Like biological systematics, ontological systematics finds genuine complexity in the basic structure of the world around us.

Notes

1. On conceptions of biological systematics, see Mayr and Ashlock 1991 and Lincoln, Boxshall, and Clark 1998. The definition is from Mayr and Ashlock, p. 2. The conception of ontology as a formal scheme of "systematics" covering all possible types of entities – what I am calling "ontological systematics" as opposed to biological systematics – derives from discussions with Chuck Dement in the Ontek research program. This conception of systematics in ontology is also informed by a close study of Whitehead's ontology, as discussed in this chapter and in Smith 2001a, reprinted in Chapter 7 as "Consciousness and Actuality."

2. See Mayr and Ashlock 1991: 430.

3. Quine argued initially against the distinction between analytic and synthetic truths, those determined by meaning and those determined by empirical fact. Subsequently, he characterized the "web of belief" as a network tied to empirical observation at the periphery, with mathematical and logical beliefs at the center. Observations are largely a posteriori, while logical-mathematical

claims are largely a priori. It is this distinction of role in the formation of beliefs that I have in mind. Categorial ontology lies at the center of our web of belief, yet is responsible to the periphery of perceptual observation.

4. See Quine 1992/1990, 1995; Smith 1994.

5. A crisp summation of received naturalist wisdom is found in Kim 1998. Kim finds room for mind as supervenient on basic physical processes that are ultimately functional states of a physical system composed ultimately of physical particles.

6. We follow the text in the Irwin and Fine (1995) translation of Aristotle's *Categories*. Fine and Irwin firmly read Aristotle as concerned with ontological, not grammatical, categories. On Aristotle's term *kategoriai*, or "predications," and thus the derived term "categories," see Irwin and Fine's comments in Aristotle 1995: notes, pp. 1, 606–7.

7. A perceptive contemporary philosophical study of Aristotle's full metaphysics, very congenial to our concerns, is Furth 1988.

8. See Furth 1988.

9. Descartes uses the term "mode" to mean a property in a substance, and "attribute" in a more special sense. Here I substitute attribute for mode in summarizing Descartes's ontology, to indicate variation on the Aristotelian system and to make room here for my own use of mode.

10. My colleague Alan Nelson reads Descartes in a way that I would place squarely between Aristotelian and Husserlian doctrines of attribute. For Descartes, on Nelson's reading, every "mode," or instance of one of the attributes of thought and extension, is itself abstracted from a substance only by a specific mental act of "distinction," so that, as I understand the view, each mode of thinking-such-and-such or being-extended-thus-and-so is inseparable from its substance except by this rational act of distinguishing it in a certain way. Not only is Descartes beginning to observe the formal-material distinction in ontology; he is also beginning to articulate a notion of modes as inseparably dependent elements in proper substances. Both views emerged in full flower in Husserl's ontology, as we see later. Nelson's reading of Descartes, as yet unpublished, was presented in Paul Weingartner's seminar on ontological categories at the University of California, Irvine, in spring 2001.

11. The similarities and differences between Husserl and Wittgenstein on these issues are traced out in Smith 2002c.

12. See Smith 2002c. See Tougas 1996 on Wittgenstein on the "projection" of states of affairs.

13. For details see Smith 2002b.

14. See Smith 2002a, 2002d.

15. See the details in Smith 2002b.

16. See Smith and McIntyre 1982, on the structure of intentionality according to this analysis. The details make the case glossed here that intentionality cannot be a proper relation, for reasons here assimilated to formal ontology.

References

Aristotle. 1995. *Categories*. In Aristotle, *Selections*. Translated by Terence Irwin and Gail Fine. Indianapolis: Hackett.

Armstrong, David M. 1996. *A World of States of Affairs.* Cambridge: Cambridge University Press.

Chalmers, David. 1996. *The Conscious Mind.* Oxford: Oxford University Press.

Descartes, René. 1984. *Meditations on First Philosophy.* In *The Philosophical Writings of Descartes,* vol. 2, translated by John Cottingham, Robert Stoothoff, and Dugald Murdoch. Cambridge: Cambridge University Press. Latin original, 1641.

Furth, Montgomery. 1988. *Substance, Form and Psyche: An Aristotelian Metaphysics.* Cambridge: Cambridge University Press.

Husserl, Edmund. 1963. *Ideas.* Translated by Boyce Gibson. New York: Collier Books. From the German original of 1913. Originally titled *Ideas pertaining to a Pure Phenomenology and to a Phenomenological Philosophy, First Book.*

2001. *Logical Investigations.* Vols. 1 and 2. New edition, edited with an introduction by Dermot Moran, and with a preface by Michael Dummett. London: Routledge. Translated by J. N. Findlay from revised, second German edition. London: Routledge and Kegan Paul, 1970. German original, 1900–1, revised 1913 (Prolegomena and Investigations I–V), 1920 (Investigation VI).

Kant, Immanuel. 1997. *Critique of Pure Reason.* Translated and edited by Paul Guyer and Allen W. Wood. Cambridge: Cambridge University Press. German original, 1781/1787.

Kim, Jaegwon. 1998. *Mind in a Physical World.* Cambridge, Mass.: MIT Press.

Lincoln, Roger, Geoff Boxshall, and Paul Clark, eds. 1998. *A Dictionary of Ecology, Evolution and Systematics.* 2nd ed. Cambridge: Cambridge University Press.

Mayr, Ernst, and Peter D. Ashlock. 1991. *Principles of Systematic Zoology.* 2nd ed. New York: McGraw-Hill.

Quine, W. V. 1992. *Pursuit of Truth.* Rev. ed. Cambridge, Mass.: Harvard University Press. First edition, 1990.

1995. *From Stimulus to Science.* Cambridge, Mass.: Harvard University Press.

Simons, Peter. 1994–95. "New Categories for Formal Ontology." *Grazer Philosophische Studien* 49: 77–99.

Smith, David Woodruff. 1994. "How to Husserl a Quine – and a Heidegger Too." *Synthese* 98 (1): 153–73.

1995. "Mind and Body." In Barry Smith and David Woodruff Smith, eds., *The Cambridge Companion to Husserl,* pp. 323–93. Cambridge: Cambridge University Press.

1997. "Being and Basis." Unpublished manuscript, extending my presentation under the same title at the Second European Congress of Analytic Philosophy, Leeds, England, 5–7 August 1996.

1999. "Background Ideas." Appeared in Italian translation as "Idee di sfondo," *Paradigmi (Estratto da PARADIGMI, Rivista di critica filosofica)* (Rome), Anno XVII, no. 49 (January–April): 7–37.

2001a. "Consciousness and Actuality in Whiteheadian Ontology." In Liliana Albertazzi, ed., *The Dawn of Cognitive Science: Early European Contributors,* pp. 269–97. Dordrecht: Kluwer Academic Publishers.

2001b. "Three Facets of Consciousness." *Axiomathes* 12: 55–85.

2002a. "What Is 'Logical' in Husserl's *Logical Investigations?* The Copenhagen Interpretation." In Dan Zahavi and Frederik Stjernfelt, eds., *100 Years of Phenomenology: Husserl's Logical Investigations Revisited,* pp. 51–65. Dordrecht: Kluwer Academic Publishers.

2002b. "Mathematical Form in the World." *Philosophia Mathematica* 10 (3): 102–29.

2002c. "Intentionality and Picturing: Early Husserl *vis-à-vis* Early Wittgenstein." *Southern Journal of Philosophy* 40, suppl.: 153–80.

2002d. "'Pure' Logic, Ontology, and Phenomenology." *Revue internationale de philosophie.*

Smith, D. W., and R. McIntyre. 1982. *Husserl and Intentionality.* Dordrecht: D. Reidel.

Tougas, Joseph. 1996. "Hertz und Wittgenstein. Zum historischen Hintergrund des Tractatus." *Conceptus* 29, 75: 205–28.

Whitehead, Alfred North. 1978. *Process and Reality.* Edited by David Ray Griffin and Donald W. Sherburne. New York: Free Press, Macmillan. First edition, 1929.

Wittgenstein, Ludwig. 1961. *Tractatus Logico-Philosophicus.* Translated by D. F. Pears and B. F. McGuinness. London: Routledge and Kegan Paul. German original, 1921.

Coda: The Beetle in the Box

I found a little beetle, so that Beetle was his name,
And I called him Alexander and he answered just the same.
I put him in a match-box, and I kept him all the day . . .
And Nanny let my beetle out –
. . . And Beetle ran away.
. . . .
We went to all the places which a beetle might be near,
And we made the sort of noises which a beetle likes to hear,
And I saw a kind of something . . .
It was Alexander Beetle I'm as certain as can be
And he had a sort of look as if he thought it must be ME.

A. A. Milne, "Forgiven"

Now someone tells me that *he* knows what pain is only from his own case! –
Suppose everyone had a box with something in it: we call it a "beetle." No
one can look into anyone else's box, and everyone says he knows what a
beetle is only by looking at *his* beetle. – Here it would be quite possible for
everyone to have something different in his box. One might even imagine
such a thing constantly changing. – But suppose the word "beetle" had a
use in these people's language? – If so it would not be used as the name of
a thing. The thing in the box has no place in the language-game at all; not
even as a *something*: for the box might even be empty. – No, one can "divide
through" by the thing in the box; it cancels out, whatever it is.

That is to say: if we construe the grammar of the expression of sensation on the model of "object and designation" the object drops out of
consideration as irrelevant.

Ludwig Wittgenstein, *Philosophical Investigations*

Friends of Pooh will know this poem. Friends of another teacher of children (who moved from Austria to England in Pooh's time) will know

the prose passage about a differently boxed beetle. (Would the serious Austrian philosopher read English children's poems? "Lewis Carroll" is one of the few proper names in the index of the *Investigations*.)

With his play of the beetle in the box, some think, the Bard of Oxbridge may have undercut (deconstructed?) the very discipline of phenomenology. I think not, for reasons emerging in the chapters in this volume, and accordingly I should like to remark briefly, in closing this book, on our familiar background picture of contents in the "box" of consciousness.

Wittgenstein continued, through the *Investigations*, to ponder questions of consciousness, privacy, introspection, and their role (if any) in language – in ways not altogether uncongenial (surprise?) to the present studies. Contrasting the following questions

"Are these books *my* books?"
"Is this foot *my* foot?"
"Is this body *my* body?"
"Is this sensation *my* sensation?"

Wittgenstein asked, "Which sensation does one mean by '*this*' one? That is: how is one using the demonstrative pronoun here? Certainly otherwise than in, say, the first example! Here confusion occurs because one imagines that by directing one's attention to a sensation one is pointing to it" (1997/1953: §411). I find no such confusion in the careful practice of phenomenology.

We do direct our attention to our own experiences of sensation, perception, thought, volition: we do so in *reflection* in the practice of phenomenology. And we may aptly use demonstrative pronouns in phenomenological analysis, as we have in the chapters in this volume: "In *this* very experience I see this frog." When I use this sentence to articulate the structure of my own experience, the second occurrence of the demonstrative (in "this frog") articulates my perceptual intention of an object, whereas the first occurrence of the demonstrative (in "in this experience") articulates my inner awareness of my perceptual experience. Notice that there is here no higher-order observation accompanying my experience, wherein I observe "*this* experience," as if mentally pointing to it while I turn my attention to it, with you-my-reader as if peering over my mental shoulder as I mentally point. Nor am I using the pronoun "this" while I demonstrate my experience, say, by pointing to the top of my head (indicating something happening in my visual cortex an inch below my fingertip). No, carrying out the phenomenological analysis is not like that. ("What is it like?") It is not like looking and pointing at the contents in one's own private box of consciousness. Analysis of experience is rather more like attending to what you are saying and then reflecting on its form and meaning. Thus, we have all experienced perceptions like that exampled. We may call upon our familiarity with such experiences (by virtue of inner awareness) and our recollections of such experiences. And we then try to *describe, interpret,* and *analyze* the form and meaning or content of such experiences. One technique of explication and analysis is to formulate a phenomenological description of the form: "In this experience I see this frog." In using this sentence I am relying on my readers' familiarity with that type of experience, and I am using that form of sentence to articulate that familiar form of consciousness, including the form of inner awareness of the experience. The first-person methodology here spreads from the singular to the plural: I am talking to *you*. Similarly, Marcel Proust's first-person descriptions of his own relived experiences, in *In Search of Lost Time*, invite the reader to explore those types of experience as if they were the reader's own. But we readers don't think Proust is pointing his finger at his remembered experiences!

There is no confusion in either use of the demonstrative pronoun in the phenomenological description, "In this experience I see this frog." Indeed, a number of philosophers have by now used the logic or grammar of demonstratives in the analysis of consciousness, assuaging Wittgensteinian worries about our inner ostension of our own mental states or their contents. (If we can't point at them, we can't talk about them? And we can't direct our attention to them either?)

After Jaakko Hintikka (1962, 1969) developed a modal logic of (sentences ascribing) intentional attitudes including perception, Hector-Neri Castañeda (1966, 1968) laid out an analysis of the use of "indicator" and "quasi-indicator" pronouns in attributions of belief or perception. Thus, when Hector says, "David thinks that he himself is a phenomenologist," Hector ascribes to David a thought David would express by saying, "I [myself] am a phenomenologist." And when Hector says, "David sees that this* frog [before David himself] is jumping," Hector ascribes to David a visual judgment David would express by saying, "This frog [before me myself] is jumping." Castañeda's aim was partly to resolve the quandary of "private" language about mental states, by showing how our language allows us to attribute to others their own "private" perspectives, in ways that are already familiar in ordinary language (regimented a bit for logical purposes). Indeed, half a century earlier, Husserl had already begun the appraisal of the use of "this" to express the content of one's perception in saying "This blackbird flies up" (*Logical Investigations*, 1900–1; I, §26, V, §5.).

Nodding to Castañeda, John Perry (2001) has recently studied "reflexive" thought contents, using "inner demonstratives" to indicate contents of one's own thoughts whose truth conditions are tied to the mental state in which they occur. And nodding to Perry and Castañeda (as well as Wittgenstein), while drawing on cognitive studies of somatic proprioception, José Luis Bermúdez (1998: 146) has used demonstratives to characterize different types of self-consciousness, both bodily and psychological.

Meanwhile, looking to phenomenological analyses begun by Brentano, Husserl, and Sartre, I have been using inner demonstratives (as here) to articulate the formal structure of one's inner awareness of one's own passing experience: "In *this* experience I see this frog." (The background of this line of analysis is found in Smith 1986, Smith 1989, and Smith and McIntyre 1982, with roots in the original debate between Hintikka and Castañeda.)

If these post-Wittgensteinian views of consciousness are valid, then it remains perfectly appropriate to speak of experience as "inner" and our immediate awareness thereof as "inner." The question is what that amounts to, and the chapters in this volume pursue just that question. Indeed, we should now see that Wittgenstein's worries about the "beetle in the box" concern how our public language about experience works, and not whether we have conscious experience, which Wittgenstein pointedly affirms (*Investigations*, §§420, 423). Yes, there is a beetle in the box, and we can talk about it: not by saying "this" while pointing a finger as if toward a bug in one's hair or head, but by using language in ways diagnosed by Castañeda et alii, and by expressing phenomenological structures of experience appropriately.

Indeed, Wittgenstein's own sensibility about consciousness resonates eerily in contemporary philosophy of mind:

> The feeling of an unbridgeable gulf between consciousness and brain-process: how does it come about that this does not come into the considerations of our ordinary life? . . . THIS is supposed to be produced by a process in the brain! – as it were clutching my forehead. – But what can it mean to speak of "turning my attention on to my own consciousness"? (*Investigations*, §412)

The mind-body problem, incorporating (in-corp-orating) today's "hard" problem of consciousness, remains puzzling even as we both embrace neuroscience and turn in phenomenological reflection to the structure of consciousness as experienced. This combination of phenomenology and natural science occupied our attention in the preceding chapters. And what it means to turn my attention to my own consciousness, and to its typical structure, is: to practice phenomenology, as it has grown out of its roots in Husserl's work and as it has evolved in contemporary modes of analysis. (Indeed, attention itself has a complex phenomenological structure that commands our attention in phenomenology: see Gurwitsch 1985 and Smith 2003.)

There is a "picture" of consciousness that we undeniably and *correctly* hold, as Wittgenstein explicitly stressed (*Investigations*, §§422–24). It is the centerpiece of The Picture, the cartoon early in the present book. That picture is a part of our background world picture: in precisely the sense appraised in the essay in Chapter 5 called "Background Ideas," which drew on Wittgenstein's conception of "background" (in *On Certainty*, 1972/1949–52). So familiar is the picture that our everyday cartoon form canonizes it. (And when did the cartoon idiom develop? How long after the Cro-Magnon cave drawings?) That picture is a guide to the phenomeno-onto-logical structures explored in these chapters. There we find "the beetle in the box": the content in an act of consciousness in the surrounding world of nature and culture and language.

References

Bermúdez, José Luis. 1998. *The Paradox of Self-Consciousness.* Cambridge, Mass.: MIT Press.

Castañeda, Hector-Neri. 1966. " 'He': A Study in the Logic of Self-Consciousness." *Ratio* 8: 130–57.

———. 1968. "On the Phenomeno-Logic of the I." In *Akten des XIX Internationalen Kongresses für Philosophie*, 260–66. Reprinted in Q. Cassam, ed., *Self-Knowledge*. Oxford: Oxford University Press, 1994.

Gurwitsch, Aron. 1985. *Marginal Consciousness.* Edited by Lester Embree. Athens: Ohio University Press.

Hintikka, Jaakko. 1962. *Knowledge and Belief.* Ithaca, N.Y.: Cornell University Press.

———. 1969. *Models for Modalities.* Dordrecht: D. Reidel.

Husserl, Edmund. 2001. *Logical Investigations.* Vols. 1 and 2. New edition, edited with an introduction by Dermot Moran, and with a preface by Michael Dummett. London: Routledge. Translated by J. N. Findlay from the revised, second German edition. London: Routledge and Kegan Paul, 1970. German original, 1900–1, revised 1913 (Prolegomena and Investigations I–V), 1920 (Investigation VI).

Milne, A. A. 1978. *Now We Are Six.* New York: Dell. First edition, 1927.

Perry, John. 2001. *Knowledge, Possibility and Consciousness.* Cambridge, Mass.: MIT Press.

Smith, David Woodruff. 1986. "The Structure of (Self-)Consciousness." *Topoi* 5 (2): 149–56.

1989. *The Circle of Acquaintance: Perception, Consciousness, and Empathy.* Dordrecht: Kluwer Academic Publishers.

2003. "The Formal Structure of Context and Context-Awareness." In Lester Embree, ed., *Gurwitsch's Relevance for the Cognitive Sciences.* Dordrecht: Kluwer Academic Publishers.

Smith, David Woodruff, and Ronald McIntyre. 1982. *Husserl and Intentionality.* Dordrecht: D. Reidel.

Wittgenstein, Ludwig. 1972. *On Certainty.* Edited by G. E. M. Anscombe and G. H. von Wright. Translated by Denis Paul and G. E. M. Anscombe. New York: Harper and Row. First edition, 1969; written 1949–51.

1997. *Philosophical Investigations.* Translated by G. E. M. Anscombe. Oxford: Blackwell. First edition, 1953.

Appendix: Background Conceptions of Ontology, Phenomenology, Philosophy of Mind, and Historical Philosophy

In the following remarks I try to outline the broad conceptual, method-ological, and historical background assumed in the preceding chapters. This background is part of the unity of views developing in the chapters themselves, although the chapters are intended to be accessible without digging into this background. Of course, one's own background is often the most difficult part of one's philosophy to articulate.

The Theory of Consciousness and World in Recent Philosophy

The headwinds in philosophy are strong. They are the winds of fashion.

Metaphysics has been under suspicion for three centuries, first in the age of science and recently in the era of cultural interpretation. Yet what we need in philosophy today is a more systematic ontology, framing a wide metaphysics that is cognizant of discoveries in physics, neuroscience, and evolutionary biology alongside analyses of our own consciousness and our cultural practices and institutions.

Phenomenology has been under suspicion for half a century, as phi-losophy turned away from conscious experience. The tradition of ana-lytic philosophy turned toward language – ordinary language, symbolic logic, then computer languages – and then toward naturalism, eyeing the marvels of quantum mechanics, relativity theory, and evolutionary biology. Continental philosophy too came to stress language, discourse, text, and cultural practice over consciousness. Yet what we need today, to understand the phenomena of consciousness much discussed in re-cent philosophy of mind oriented to cognitive neuroscience, is a more systematic phenomenology, framing a careful analysis of structures of subjective experience.

The yearning for these perspectives is all around us. Famous scientists wax philosophical in popular surveys of quantum physics, cosmology, neuroscience, evolutionary biology, dynamical systems ("chaos"). Cognitive scientists, especially philosophers of mind, have rediscovered the importance of consciousness, the hard problem for once promising functionalist models of mind in computers and human brains.

To deal with the basic forms of reality, philosophy must return to systematic metaphysics or ontology. The denunciations of metaphysics – from Kant to the positivists – were aimed at speculative metaphysical systems. Yet we need a sense of system that outruns any direct experimentation. Theoretical physics follows this pattern while explaining results of empirical investigation. As W. V. Quine has well observed, our belief system – from everyday common sense to current physics – crafts the best overall theory we can muster in the face of our overall empirical observations. What we have lost in recent philosophy is the really big picture, writ large but written carefully. First there was Rudolf Carnap's logical empiricism of the 1930s and 1940s, where the new logic was to formulate our scientific theories based on reports of sensory perception (formulated in "protocol" sentences): a Kantian world view written in a language of symbolic logic (the noumena outrun that language, so we don't really talk about them). Then there came ordinary language philosophy in the 1950s, which eschewed metaphysics. P. F. Strawson's *Individuals* (1960) then allowed "descriptive" metaphysics, which would analyze the world but only as described in ordinary language. Then, since the 1970s, came the computer revolution: computer languages tell all, from the nature of mind (artificial intelligence) to DNA reproduction to molecular combination and even molecular computers, as if all reality is constituted by algorithms in some appropriate language of computation. Meanwhile, in the continental tradition Jacques Derrida transformed phenomenology into subtle interpretations of linguistic practices or texts, seeming to jettison not only consciousness but all reality beyond the text. Other French poststructuralists followed similar lines. The most extreme antimetaphysical rhetoric may lie in Jean Baudrillard's recent notion of "hyperreality," where reality is lost in virtual reality, in the commodity of computer information that increasingly defines social and ultimately political force, displacing natural languages and the consciousness that phenomenology said underlies language. The computer model has thus displaced not only classical theories of mind but social theories as well.

To deal with consciousness once again, philosophy must return to the discipline of phenomenology. A century ago consciousness was richly studied: from Franz Brentano's *Psychology from an Empirical Standpoint* (1874) to William James's *The Principles of Psychology* (1891) to Edmund Husserl's *Logical Investigations* (1900–1) and *Ideas* (1912–13) to the literary practice of interpreting one's own experience in Marcel Proust's *In Search of Lost Time* (*À la recherche du temps perdu*, beginning in 1913). Phenomenology then flourished in the continental tradition following Husserl: in Martin Heidegger, Jean-Paul Sartre, Maurice Merleau-Ponty, Simone de Beauvoir (even as Heidegger had resisted talk of "consciousness"). But analytic philosophy of mind took an apparently antiphenomenological turn with Gilbert Ryle's *The Concept of Mind* (1949), declaring that mind is but a disposition to overt behavior (ascribed in ordinary language). (In fact, Ryle viewed his own work as a variant practice of phenomenology, as did Wittgenstein his.) There followed the seminal essays of today's materialism by U. T. Place, J. J. Smart, and D. M. Armstrong in the 1950s and 1960s, positing that mind is identical with states of the central nervous system. By 1970 functionalism set in as a refinement: mental states are functional states of a physical system – mind is what the brain does in its physical environment of body in nature, and what computers can in principle do. Gradually we witnessed the return of consciousness. From Thomas Nagel's "What Is It Like to Be a Bat?" (1973) to John Searle's *The Rediscovery of the Mind* (1993), philosophers began to insist that consciousness is real even though realized in the brain in a physical universe. Specifically, consciousness has essential properties, including intentionality and sensory qualia, that do not reduce to physical properties of physical particles or even functional properties of systems of physical particles. Thus was born, or reborn, the "hard" problem of consciousness, as David Chalmers dubbed it in *The Conscious Mind* (1996). Gradually the words "intentionality" and "phenomenological" have entered the vocabulary of cognitive science – as scientific philosophy of mind has been called since the 1970s. And so, merging the rich continental and analytic traditions in philosophy of mind, we do philosophy in the new millennium with our eyes on consciousness. To develop a proper theory of consciousness, I submit, we must think through the basic results of classical phenomenology, starting with Husserl's monumental work. These classical results (in the mathematician's sense of achieved results) are not yet fully articulate in the analytic tradition (in the writings of Jerry Fodor, Daniel Dennett, Paul and Patricia Churchland, etc.), where the important details of phenomenological analysis are largely glossed over. Yet we must proceed in phenomenology today with our eyes on the issues of mind-in-brain that are so sharply focused in recent cognitive science.

A philosopher well versed in both continental and analytic philosophy observed to me recently, "It will take two hundred years for philosophy of mind to catch up with phenomenology." Indeed, the distinctions and conceptualizations of phenomenology are well ahead of the literature of analytic philosophy of mind. On the other hand, these results must today be worked into the best results of contemporary science. And one day soon results of neuroscience will help to refine results of phenomenology, as we learn where and how key structures of consciousness are produced or realized in particular patterns of neural activity in various parts of the brain.

To deal with this bewildering complexity, from quarks to consciousness to human language to computational systems to biological processes, philosophy needs a proper synthesis of phenomenology and ontology, a framework able to combine the results of phenomenology on consciousness and the results of ontology on forms of things in the world. Although doctrines of functionalism and supervenience have stimulated a great deal of thought about the precise ontological form of mind, there is more to ontology than meets our eyes in this piecemeal approach to the mind-body problem – as exciting and fruitful as the recent debates in philosophy of mind have been. It was Wilfrid Sellars who described philosophy's task as seeing together, stereoscopically, the scientific and manifest images of man. We may say instead: seeing in unity the ontological and phenomenological aspects of the world.

A Shifting Paradigm

We awake and see and hear and smell the things of morning, and the rest of waking life, and we experience variations on these things when dreaming. We perceive, think, feel emotion, value, will, move intentionally, act in meaningful ways, and interact socially, with varying degrees of consciousness. We live among other sentient beings, other subjects, notably other humans in a highly evolved social context of family and friends, houses and carpenters, banks and economists, governments and politicians. And all this occurs in a world we increasingly understand as founded on elementary particles like quarks and electrons in fields of gravity, electromagnetism, and quantum superpositions in the cosmos. We humans, blessed and afflicted with the brain of *Homo sapiens sapiens*, have developed ways of studying these phenomena: the physical, the experiential, the cultural. But our disciplines think in significantly different ways, and we are as yet unable to unify our divergent views of matter, mind, and culture.

In the 1930s, as contemporary physics came together, Husserl diagnosed a "crisis" in intellectual history.[1] With the development of mathematical physics since Galileo, he observed, we have increasingly "mathematized" nature, reducing its essence to mathematical forms that distance the world of nature so described from the world of everyday life, the "life world" as Husserl called it. We no longer understand how to put these two "worlds" (world concepts) together. In particular, we have produced a "paradox of subjectivity": our intentional activities in science have crafted our mathematical models of physical reality, but our subjective activities of consciousness, which produce our scientific theory, find no place in those models. Indeed, the concepts of "subject" and "object," Husserl found, no longer fit together, and this is the culmination of Kantian theory. Had Husserl lived another half century, he would have seen a more precise statement of this "crisis." Today we have "mathematized" the mind, declaring with cognitive science that intentional activities of thinking, perceiving, and the like are processes of computation, described mathematically by algorithms written in formal computer languages. But today some philosophers have come to realize that consciousness and, indeed, intentionality do not fit into this model. We face a profound scientific-philosophical crisis. We simply do not understand how to put consciousness and the physical world together in a unified view. More broadly, we do not know how to fit phenomenology together with the ontology of physical science.

To resolve this paradox in our understanding of consciousness and world, we need to synthesize ontology and phenomenology. But this synthesis of the ontological and the phenomenological requires a paradigm shift in our thinking about both consciousness and the world in general: a shift away from choosing among dualism, physicalism, idealism, cultural-relativism, and so forth and toward seeing consciousness at home in a world of nature, culture, and consciousness. The current paradigm of naturalistic philosophy, grounded in philosophy of science, holds that the world is composed of physical particles in physical fields; mental and cultural activities must then fit this mold. But consciousness and its interesting properties – intentionality, inner awareness, sensory qualia – simply do not appear in our best naturalistic description of causal-computational processes of "brain mechanics."

Only gradually have I come to realize that the integration of the phe-nomenological and the ontological requires a genuine paradigm shift, somewhat like the shifts Thomas Kuhn described in the history of science. It has become increasingly clear that the debates in philosophy of mind about materialism, functionalism, property-dualism, mind-as-brain, and mind-as-computation were missing both the rich structure of intentional-ity and the deep structure of ontological categories (or modes of being). To see these phenomenological and ontological structures in one view is the shift in perspective I have in mind. One part of the view takes in the details of phenomenology while another part of the view takes in the wide field of ontology. The best metaphor is not seeing with two disciplinary eyes but rather seeing as a hawk does: seeing simultaneously both its prey and the wider terrain, through different mechanisms in the eyes.

Kant proclaimed a "Copernican revolution" in eighteenth-century phi-losophy. Copernicus had said the Earth moves around the sun, rather than the sun around the Earth. Kant then said the objective world of "phenomena" – things as they appear, even things as described by physics – rests on our subjective activities (as the necessary conditions of the possibility of such a world). For Kant, the world spins around us insofar as our representations of the world are necessarily conditioned by the structures of our own cognitive faculties of sensation and con-ceptualization (today we stress language). Kant's revolution was thus a philosophical counterrevolution. The problem of subjectivity and objec-tivity has been with us ever since, either displacing or superimposing itself atop the problem of mind and body that Descartes had defined in the seventeenth century. I believe the proper perspective has emerged only with a correct account of intentionality, beginning with Husserl's ground-breaking work around 1900. But rather than say that consciousness spins around physical particles, or alternatively that the physical spins around mind or language, or that only the physical or the mental or even the cultural exists – rather, we need a new way of seeing things, a unified story that puts these things in their place while giving each their due.

Whereas Kant chose the term "revolution" (in the time of the American and French revolutions), Kuhn in *The Structure of Scientific Revolutions* (1962, a time of cultural revolution) coined the term "paradigm shift" in reviewing the major shifts in science, as from Newtonian physics to rel-ativistic and quantum physics. With such a shift, we do not simply revolt against a prior view of things; rather we come to see things in a new way that renders the old way ineffective. In philosophy, as opposed to science, a major change of views is rarely a bona fide "revolution" because the problematic phenomena – say, consciousness, nature, culture, and their interrelations – do not go away but come to be viewed differently. "Evolution" rather than "revolution" is the way of deep philosophical thinking and often scientific thinking, too (and even political thinking). Our brain type did not displace the lizard brain type; it grew from it, evolving radically new abilities but retaining many of the older, well adapted features. And so it is with philosophical theory or understanding. We move from Anaximander to Plato and Aristotle and on through Avicenna and Ockham to Suarez and Descartes to Kant to Husserl to Quine and on to our own best views that grow from prior ideas. The shifting paradigm of consciousness, bringing the phenomenological into the ontological and vice versa, follows this pattern of evolution, moving from one species of view to another about our awareness of the world in which we find ourselves.

A recurring approach to the mind-body or subject-object problem is to say that proper background assumptions prevent the problem from arising. It is said that Heidegger, Merleau-Ponty, Wittgenstein, and others – perhaps Kant or Husserl before them – gave accounts of our activities in the world, on which the old bugbears of dualism, realism, idealism, and the like simply do not arise. But in fact, after we have given careful descriptions of our experiences, practices, and language games, after we have a careful phenomenological analysis of our conscious intentional activities (and related social-cultural activities), we must extend our story with the compelling results of today's physics, biology, neuroscience; these are not just another "story" for our entertainment. And we find that phenomenological descriptions of our activities of consciousness do not fit with advancing theory in physical science, with descriptions of activities from quarks to neurons and DNA – much as Descartes had begun to see in the seventeenth century. But the ontological maneuvers of dualism, materialism, idealism, antirealism, functionalism, computationalism, connectionism, externalism, evolutionism – so ably explored in recent decades in philosophy of mind and cognitive science – do not solve, resolve, or dissolve the problem of fit. The subjectivity and intentionality of consciousness remain just that, not something else like a pattern of particles bobbing in the void.

We might tackle directly the details of physics, biology, and neuroscience, hoping to find consciousness amid the details. This tack is taken by many, from physicists like Roger Penrose to biologists like Edward O. Wilson to philosophers like David Albert, Paul Churchland, Patricia Churchland, Daniel Dennett, Fred Dretske, Jerry Fodor, Ruth Millikan, and others, all seeking to "naturalize" consciousness in one way or another. Alternatively, I urge, we might tackle directly the details of phenomenology and ontology, which must be integrated as a basis for bringing into the picture the results of recent physical-biological-computational science. Here in philosophy lies the "formal" side of the synthesis; there in empirical science lies the "material" side. And the distinction between "formal" and "material" ontology is itself part of ontology.

So we need a new way of looking at consciousness and its place in the world. We need a systematic integration of the phenomenological and the ontological (not to be confused with the mental and the physical). The yearning for this new synthesis, we noted, is all around us. Evidently the paradigm shift is in progress. However, the core of the synthesis lies in principles of philosophy, in the integration of two often misunderstood philosophical disciplines.

Of Ontology and Phenomenology

Ontology is the science or theory of being, of what is and how it is. I make no distinction here between ontology and metaphysics. Some philosophers have defined "metaphysics" as speculation about being, speculation beyond the range of our abilities to know, especially beyond all empirical knowledge. I do not use that definition (respect for Kant notwithstanding). As Quine has stressed with Peirce (and Husserl), all our knowledge is in some measure speculative, beyond the measure of absolute certainty, and limited to the range of our sensory and intellectual and intuitive capacities, limited even by our technologies that aid in developing knowledge. We practice ontology by saying, to the best of our current abilities, what is.

Phenomenology, I assume, is the science or theory of consciousness as we experience it – literally, the theory of "phenomena" in the ancient Greek sense of "appearances." Where ontology studies reality, phenomenology studies appearance, namely, the ways things are given in perception, thought, imagination, volition, and other processes. But "phenomena" are not simply sense data (as in the British empiricist tradition). Rather, "phenomena" are nearly always intentional presentations of things: things as given in perception, imagination, thought, desire, will, and indeed embodied intentional action. (Remember that perception is itself partly action, as we move our eyes, hands, nose in perceiving things around us.)

Phenomenology, as I understand it, is centered in the theory of intentionality via content or meaning, a theory pursued in prior works.[2] The present essays extend this basic conception of intentionality and place it in a framework of basic ontology.

By phenomenological analysis we learn that consciousness is characteristically intentional: in (nearly) each case it is of or about something, directed toward some object. And we learn that each act of consciousness "intends" its putative object through some conceptual content or meaning. So phenomenology studies the meanings things have for us in our intentional activities of perception, thought, imagination, emotion, volition, and bodily action. Moreover, our conscious intentional activities tie into attitudes, moods, skills, habits, traditions, languages, and ways of life that are part of our experience even if we are not directly aware of them. Thus, phenomenology comes down to the study of intentionality via meaning, first in consciousness and then in less-than-conscious activities that carry meaning. Historically, phenomenology as we know it was launched in 1900 with Husserl's analysis of intentionality via meaning. Husserl drew on Brentano, Bolzano, Descartes, Hume, and Kant. In Husserl's wake, Heidegger defined phenomenology as the study of "phenomena"; however, avoiding the terminology of "consciousness" and "subject," Heidegger spoke of our "being" and "comportment" (*verhalten*, relating to) in acting, speaking, and thinking. I adopt a simplified terminology of consciousness, context, subject, act, content, and object – as appraised in the present essays and in prior works.[3]

Phenomenology is a philosophical discipline alongside ontology, epistemology, logic, ethics, and the like. Yet phenomenology is still not so well understood as these other disciplines. Some philosophers have defined phenomenology as a movement, a historical tradition including the work of Husserl, Heidegger, Sartre, Merleau-Ponty – and more recently Gurwitsch, Mohanty, Føllesdal, and many others. I do not follow that definition (admiration for Husserl, Heidegger, et al. notwithstanding). Other philosophers have defined phenomenology as the practice of a particular philosophical methodology: transcendental reflection on the meaning of our experience, as prescribed by Husserl; or hermeneutic interpretation of our existence, as prescribed by Heidegger. It is wrong to restrict phenomenology to method. This is a process-product confusion: we practice phenomenology by a process of analysis following some methodology, and our results are products thereof. No, phenomenology is a discipline, the study of consciousness *cum* meaning, and in its practice we will try to settle upon the best methodology. We experience a variety of types of consciousness, we reflect upon them, we describe and interpret and analyze them as best we can, we expand on the conditions in which we believe them to arise, physical as well as cultural and environmental conditions. In short, we do the best we can. Perhaps results of neuroscience or evolutionary biology or quantum physics will one day help us to refine our phenomenological descriptions of experience as we live it. Or perhaps only disciplined practices of meditation, like those of Tibetan Buddhists or Zen masters, can refine our awareness of our own experiences. But in any event, we must describe, interpret, and analyze our experience as we find it.

Our best philosophical theory of mind-and-world will unfold within a system of phenomenology and ontology. This is nearly a tautology, given the preceding. More significantly, the two disciplines are interdependent. Ontology must accommodate the results of phenomenology, because what exists includes our acts of consciousness and their intentional structure; moreover, the practice of ontology is a collective intentional activity that we philosophers engage in. And phenomenology must accommodate the results of ontology, because an act of consciousness has an identity, properties (including intentionality), relations, and dependencies – all ontological structures of the act of consciousness.

These ties between ontology and phenomenology are apparent in the classical literature of phenomenology. Husserl famously analyzed structures of consciousness while bracketing questions of the existence of the objects of consciousness. Yet, within those phenomenological analyses, he assumed basic principles of what he would call formal ontology – for example, particular things have properties or "essences," or they are dependent beings. And Heidegger famously described our intentional activities or "comportments" such as hammering or indeed speaking as modes of "being in the world." These ontological structures of "being" are apparent in our own activities.

From a Phenomenological Point of View

When Quine titled his book of essays _From a Logical Point of View_ (1953),[4] it was timely to reassert the importance of logic in philosophy. I say reassert because logic has been very much with us ever since Plato and Aristotle. Looking back on twentieth-century philosophy, we take modern logic for granted – because of the work of Frege, Whitehead and Russell, Gödel, Carnap, Tarski, Quine, and others. Yet, we know, Russell and Moore were surrounded by the British absolute idealists, and Harvard was the heart of pragmatism. (Never mind that Peirce in Cambridge, Massachusetts, was also a logician and McTaggart in Cambridge, England, reasoned elegantly.) Quine's title is vindicated by developments throughout the second half of the century, in the traditions of philosophical logic, philosophy of language, possible-worlds semantics, the logic and metaphysics of modality, externalism of direct reference and of intentional content, and indirectly the developments in philosophy of mind under the umbrella of cognitive science.

It is now timely to reassert the importance of phenomenology in philosophy, to look afresh at mind and its place in the world "from a phenomenological point of view." I say reassert, because phenomenology has been with us incipiently in the ancient Greek and Indian thinkers, formatively in the medieval Arabic-Islamic and European-Latin philosophers, and modernly in Descartes, Hume, Kant, Fichte, Bolzano, and Brentano. Of course, phenomenology has been with us explicitly, and prominent in European thought, since 1900 when Husserl crystallized the theory of intentionality and the discipline of phenomenology as science of consciousness – our own conscious intentional experience in perceiving, thinking, and acting. It is timely now to reassert, and reassess, phenomenology in philosophy of mind and in metaphysics. For consciousness has drawn intense interest as science has begun to understand the workings of the brain in recent decades.

Why must we reassert, if you will, consciousness itself? Why must we reassert the discipline of phenomenology in the study of consciousness?

Think of philosophy's trends. Existentialism stressed consciousness; but poststructuralism forgot it while privileging structures of language. Philosophical logic in Quine, taking behaviorism as key to the study of meaning in language acquisition, seemed to bypass consciousness – though Quine himself says consciousness is not to be denied but rather is a mystery to be explained.[5] Ordinary language philosophy of the 1950s and 1960s followed later Wittgenstein's seeming suggestion that mind is in effect language and talk of "consciousness" is a muddle – although P. F. Strawson distinguished mental and physical predicates and dug into Kant. Modal logic led Hintikka into the logic of belief and perception, and intentionality in phenomenology. Føllesdal wrote evenly about Husserlian intentionality, modal logic, and Quinean linguistic meaning (and indeterminacy). Then cognitive science took the computer as model of mind, whence consciousness and even intentionality seemed to escape the theory of mind. Neuroscience spawned a different, connectionist computer model, but still consciousness remains today the chief stumbling block for a scientific, physicalistic theory of mind.

Here is our entry. We need to study consciousness itself from the first-person point of view, from the perspective of each of us who experience it. And that is precisely the business of phenomenology.

My approach to phenomenology is a syncretic approach that integrates classical phenomenology (building especially on Husserl's work) with logical-semantic theory (in the tradition of Frege, Tarski, Carnap, Hintikka, and Kaplan) – and, I would add, with recent cognitive neuroscience (in the spirit of Searle and others who take consciousness seriously), and indeed with the wider sweep of Western philosophy (from Plato and Aristotle through Ockham and Aquinas to Descartes, Hume, and Kant into Bolzano, Brentano, Husserl, and beyond). This integrative approach is sometimes called "California phenomenology."[6] There is a wink in this appellation, not least because the relevant philosophers are not confined by the shores and mountains of the Golden State.

From an Ontological Point of View

A particular type of ontology is part of what I am urging in this book. It appears in the guise of "formal ontology" and "basic ontology." I do not mean ontology written in a formal language such as the predicate calculus or set theory or the computer language of the day. Rather, formal ontology concerns "formal" structures of reality, such as the formal categories Individual, Property, State of Affairs, Number, Group, Part, Dependence, and so on, which govern very different domains such as the material categories Mind, Body, Culture. By "basic" ontology I mean the study of particularly fundamental categories, including what I like to call basic "modes of being," such as being an individual, being a part, being dependent, and being intentional.

As our perspective widens from issues of phenomenology to issues of basic ontology, our conception of consciousness and its place in the world widens. We come to see the importance of the synthesis of phenomenology and ontology. Only thereby, I believe, can we truly understand the phenomenon of mind. And contemporary theory of mind needs to approach mind systematically in this way, or so I argue in different ways in the present chapters.

In fact, we *experience* certain basic ontological structures. In perception, for instance, I experience the distinctness and relation between "that object before me" and myself, or "I." The articulation of such structures in a developed ontology lies beyond the deliverance of our inner awareness of our experience. Yet we still have a phenomenological hold on part of the structure of the world.

It is timely now to reassert the importance of ontology in phenomenology and in contemporary philosophy of mind. I say reassert, because ontology was explicitly used in the classical texts of phenomenology, in Husserl and indeed Heidegger; and because in philosophy of mind since the 1950s the mind-body problem has been attacked from positions that identify a mental state with a brain state or a functional state of the body or a social condition of the person, or from other stances that are specifically ontological. What is needed now is a more systematic approach to both the phenomenological and ontological structures of mind, especially consciousness. This metaphilosophical thesis is a corollary of the present collection of essays, which study a variety of these aspects of phenomenology and ontology, aspects that are less than systematically appraised in the rich literature on philosophy of mind in the past decades.

From a Historical Point of View

My aim in this book is a philosophical account of aspects of consciousness and world. My technique is partly what may be called *historical* philosophy. By this I mean that I draw philosophical analysis partly from historically significant works, relying on interpretation of key texts along the way, although my aim is neither exclusively nor primarily "historical."

Many philosophers today seem to think that philosophy is done in the absence of the voices of Plato, Aristotle, Descartes, Hume, Kant, Husserl, Russell, Quine. "To the arguments themselves, to the concepts and theories themselves, to the phenomena themselves," we cry. Indeed. But the best philosophical ideas we have nearly always begin their lives clothed in texts of the great thinkers of the past, and often we must go back and work at understanding classical texts in order to move on with the theories and arguments themselves. Our philosophical theories are historical products, and much of their real content lies in their evolution, in debates along the way. Most important, in the ontology and phenomenology of consciousness, the best groundwork is laid not in the current works of Fodor, Dennett, and others (as exciting as these are), but in penetrating investigations by Descartes, Kant, Brentano, Husserl, Sartre. To make real progress, we must develop new conceptual tools sharpened by issues of neuroscience, computer science, evolutionary theory, social theory, and other disciplines. No one I know has a full mastery of all these things, so the work will be collective.

To understand a philosophical idea, such as the concept of intentionality or indeed of consciousness, we need to address it in texts that speak to us (in our style and cultural context), in precise formulations (in exact theories and arguments), and in historical genesis (in dialectic that develops or reveals motivations). These three facets of an idea are instances of what I have called its appearance, form, and substrate (in the essay in Chapter 1). For reasons of *ontology*, then, the content of a philosophical idea lies partly in its history!

At various points along the way I have drawn on historical philosophical works, by Aristotle, Descartes, Husserl, Heidegger, Wittgenstein, Whitehead, Quine, and others. I do not see the history of philosophy as a dustbin of old and dead ideas. Rather, I see our current ideas as evolutionary heirs of the ideas and debates of years past, whether last year or 50 years ago or 300 years ago – or even 2,600 years ago (as I like to nod to Anaximander at the dawn of philosophy as we know it).

To the Things Themselves

With these metaphilosophical and methodological remarks in the background, drawing on "background ideas" of our philosophical culture, the present chapters go about their business, looking to consciousness and its place in the world.

Notes

1. See Edmund Husserl, *The Crisis of European Sciences and Transcendental Phenomenology*, trans. David Carr (Evanston, Ill.: Northwestern University Press, 1970; from the posthumous German edition, 1954; original German texts from 1935–38). See especially §§53–54 on "the paradox of subjectivity."
2. This conception of intentionality was detailed in David Woodruff Smith and Ronald McIntyre, *Husserl and Intentionality* (Dordrecht: D. Reidel, 1982), and ramified in David Woodruff Smith, *The Circle of Acquaintance* (Dordrecht: Kluwer Academic Publishers, 1989). A wider ontology for the basic theory I reconstructed from Husserlian principles in "Mind and Body," in Barry Smith and David Woodruff Smith, eds., *The Cambridge Companion to Husserl* (Cambridge: Cambridge University Press, 1995), pp. 323–93.
3. See Smith and McIntyre, *Husserl and Intentionality* (1982, cited above), and D. W. Smith, *The Circle of Acquaintance* (1989, cited above).
4. See W. V. Quine, *From a Logical Point of View* (Cambridge, Mass.: Harvard University Press, 1953, 1961, 1980). The origin of Quine's title was in fact

Harry Belafonte's performance of the calypso melody "From a Logical Point of View"; see the foreword to the 1980 edition.

5. See W. V. Quine, *Quiddities: An Intermittently Philosophical Dictionary* (Cambridge, Mass.: Harvard University Press, 1987), pp. 132–33.

6. The tradition of California phenomenology includes works of Dagfinn Føllesdal, Jaakko Hintikka, Hubert Dreyfus, Ronald McIntyre, Izchak Miller, and the present author. Kindred work drawing on ontology includes that of Dallas Willard, Barry Smith, Kevin Mulligan, and Peter Simons. *The Cambridge Companion to Husserl* extends these literatures, including essays by J. N. Mohanty, Kit Fine, Herman Philipse, and Richard Tieszen. Mohanty's many works on phenomenology are sympathetic to analytic philosophy of logic and language, whereas philosopher-of-language Michael Dummett approaches Husserl from Frege in the logical tradition. The writings of Aron Gurwitsch, Lester Embree, Robert Sokolowski, John Drummond, and many others on phenomenology do not tie directly to the analytic tradition, yet their results in phenomenology are consonant on many points with results in California phenomenology – as we should expect if we all turn to "the facts [*Sachen*] themselves," the facts of conscious experience.

Index

a posteriori, 246
a priori, 245
abstraction, 20; *see also* form;
    formalization
acquaintance, 2
action, phenomenology of, 4, 123–124;
    awareness in, 125; awareness of, 128;
    experience of acting, 130; structure of,
    138
actual entity (Whitehead), 216, 217
actual occasion (Whitehead), 216
actual world (Whitehead), 218
actuality (Whitehead), 218
adverbial character, consciousness as, 119
agent, of embodied action, 141
ambulo ergo sum, 5
analytic philosophy, 289; *see also*
    continental philosophy
analytic truth, in the cogito, 63
analytic/synthetic distinction, as opposed
    to formal/material distinction, 183; *see
    also* formal/material distinction
Anaximander, 221, 238, 268
Anscombe, G. E. M., 73
apeiron, 221, 238, 268
appearance, 3, 17, 276
apperception (Leibniz), 78
archetypes, Jungian, 168
Aristotle, 6, 32; Aristotle's categories, 179,
    248; Aristotle's category scheme, 249;
    modernized Aristotelian category
    scheme, 253; ordered neo-Aristotelian
    category scheme, 254
Armstrong, David, 25, 34, 90, 262
Ashlock, Peter D., 19
atomism, 223

attention, 127, 135
autophenomenology, 82

background, 2, 5, 35–36, 140, 149;
    argument for, 153–156; earthquake
    example for, 171; and scientific theory,
    204, 287
background idea, 5, 151, 163; ontology of,
    167
background image (of the world), 149
basic ontology, 244, 300
becoming, ontological, 6, 34, 212, 214,
    215; temporal vs. ontological, 6, 219,
    238, 267, 268, 273
beetle in the box, 284, 286, 287
Bergson, Henri, 214
Bermúdez, José Luis, 286
blindsight, 80
bodily movement, 130
body, 139; lived or living (*Leib*), 144, 200;
    living-moving-willing, 141
body-body problem, 141
bracketing, 17; *see also* methodology (of
    phenomenology)
Bratman, Michael, 143
Brentano, Franz, 78, 85
Brough, John, 103
Burge, Tyler, 166

Calabi, Clotilde, 69
capacities, 152, 157; *see also* know-how;
    skills
Castañeda, Hector-Neri, 73, 136, 285
categorial ontology, of physical world, 184;
    as theory, 245; *see also* category; category
    scheme

305

Printed in the United States
By Bookmasters